S0-DUU-701

OXFORD MONOGRAPHS ON BIOGEOGRAPHY

Editors: W. GEORGE, A. HALLAM, AND T. C. WHITMORE

OXFORD MONOGRAPHS ON BIOGEOGRAPHY

In an area of rapid change, this series of Oxford monographs will reflect the impact on biogeographical studies of advanced techniques of data analysis. The subject is being revolutionized by radioisotope dating and pollen analysis, plate tectonics and population models, biochemical genetics and fossil ecology, cladistics and karyology, and spatial classification analyses. For both specialist and non-specialist, the Oxford Monographs on Biogeography will provide dynamic syntheses of the new developments.

1. T. C. Whitmore: *Wallace's line and plate tectonics*
2. Christopher J. Humphries and Lynne R. Parenti: *Cladistic biogeography*

CLADISTIC BIOGEOGRAPHY

CHRISTOPHER J. HUMPHRIES
Department of Botany,
British Museum (Natural History)

AND

LYNNE R. PARENTI
Department of Ichthyology,
California Academy of Sciences

CLARENDON PRESS · OXFORD
1986

Oxford University Press, Walton Street, Oxford OX2 6DP

Oxford New York Toronto
Delhi Bombay Calcutta Madras Karachi
Kuala Lumpur Singapore Hong Kong Tokyo
Nairobi Dar es Salaam Cape Town
Melbourne Auckland

and associated companies in
Beirut Berlin Ibadan Nicosia

Oxford is a trade mark of Oxford University Press

Published in the United States
by Oxford University Press, New York

British Library Cataloguing in Publication Data

Humphries, Christopher J.
Cladistic biogeography.—(Oxford monographs on biogeography)
1. Biogeography
I. Title II. Parenti, Lynne R. III. Series
574.9 QH84
ISBN 0-19-854576-2

Library of Congress Cataloging in Publication Data

Humphries, Christopher John.
Cladistic biogeography.
(Oxford monographs on biogeography)
Bibliography: p.
Includes index.
1. Cladistic analysis. 2. Biogeography.
I. Parenti, Lynne R. II. Title. III. Series.
QH83.H86 1985 575 85-9711
ISBN 0-19-854576-2

Set by Joshua Associates Limited, Oxford
Printed in Great Britain by
J. W. Arrowsmith Ltd, Bristol

PREFACE

It was well understood by biologists in the eighteenth century that different plants and animals had particular distribution patterns over the surface of the earth. Since that time literally thousands of biogeographical explanations have been published to account for them. Perhaps most important, as so forcefully pointed out by the biogeographer Leon Croizat, was the fact that many of the clear-cut distribution patterns showed distributional congruence when viewed on a broad scale. These 'generalized tracks' of distribution are so consistent in disjunct transoceanic terrestrial groups, like flowering plants and mammals, that they imply historical connections between biotas.

The analysis of the biotic components to determine hierarchical classifications of areas by cladistic techniques is the method identified in this book as cladistic biogeography. The use of this method to determine the significance of distributional congruence in relation to the recent dramatic discoveries in paleogeography and geology, particularly in the Caribbean and the Southern hemisphere, will be described to show some of the most recent views on the earth's complex history.

London C. J. H.
San Francisco L. R. P.
June 1984

ACKNOWLEDGEMENTS

We are grateful to Steven Blackmore, Peter Forey, Lance Grande, Gordon Howes, Gareth Nelson, Colin Patterson, Norman Platnick, Dick Vane-Wright, and the editors for useful comments on the manuscript. For advice we are also grateful to Nigel Stork and Peter Hammond. We thank Dinah Stevenson, Marilyn Humphries, and Loveday Hoskins for typing, and Bob Press for drawing most of the illustrations.

The following figures were reproduced by permission: 1.8 John Wiley & Sons, 1.11 and 2.7 Donn Rosen and the editors of *Systematic Zoology*, 2.4 Prentice Hall, 2.5 Addison-Wesley, 2.6 Columbia University Press, 2.10, 2.11, and 2.12 Donn Rosen, 4.2 *Journal of the Arnold Arboretum* and C. G. G. J. van Steenis 4.6 Minister of Supply and Services, Canada.

CONTENTS

Introduction xi

1 Historical biogeography 1
- 1.1 Introduction 1
- 1.2 Distribution patterns 1
- 1.3 A history of ideas 7
  - 1.3.1 Origins 7
  - 1.3.2 Habitations 9
  - 1.3.3 One history 10
  - 1.3.4 Separate histories 11
  - 1.3.5 Sclater's regions 13
- 1.4 Centres of origin 14
- 1.5 Croizat's tracks 15
- 1.6 Conclusions 19

2 Methodological developments 21
- 2.1 Introduction 21
- 2.2 Cladistics 21
- 2.3 Cladistics and biogeography 25
  - 2.3.1 The progression rule 26
  - 2.3.2 Vicariance biogeography—Rosen's method (1976) 29
  - 2.3.3 Cladistic biogeography—The method of Platnick and Nelson (1978) 31
    - 2.3.3.1 Poeciliid fish in Middle America—Rosen's example (1978) 33
    - 2.3.3.2 Ancestral species map—Wiley's method (1980, 1981) 37
    - 2.3.3.3 Component analysis 38
    - 2.3.3.4 Endemics 40
    - 2.3.3.5 Missing areas 40
    - 2.3.3.6 Widespread taxa 42
    - 2.3.3.7 Platnick's example (1981) 47
- 2.4 Conclusions 52

3 The real world 53
- 3.1 Introduction 53
- 3.2 Same pattern: different taxa 55
  - 3.2.1 Congruence 55
  - 3.2.2 General explanations 58
  - 3.2.3 Predictions 60

3.3 Different patterns: different taxa 60
3.3.1 Areas of hybrid origin 61
3.3.2 Patterns of different ages 63
3.4 Geology and the cladistic biogeographer 64
3.4.1 Cladograms of taxa 65
3.4.2 Cladograms of areas 66
3.5 Conclusions 67

4 A new view of the world 69

4.1 Introduction 69
4.2 Tropical versus amphitropical 70
4.3 Pangaea, Pacifica, or an expanding earth? 75
4.4 Historical biogeography of the southern end of the world 77
4.4.1 Darlington's view (1965) 77
4.4.2 The two South Americas 78
4.4.2.1 Patterns of taxa 79
4.4.2.2 Patterns of areas 80
4.4.2.3 Geology or age? 82
4.4.3 A composite New Zealand 83
4.5 A new view of the world 84
4.5.1 Pacifica and amphitropical distributions 84
4.5.2 Pacifica versus Pangaea: geology or age? 86
4.6 Conclusions 86

Bibliography 88

Glossary 93

Index of animal and plant names 95

General index 97

INTRODUCTION

Development of theory and methodology of biogeography has received attention from an increasing number of biologists and geologists during the past two decades. Biogeography means many things to many people depending on their outlook and purposes of enquiry. As data accumulated over the last two centuries it became clearer that organisms exhibit particular distribution patterns over the surface of the earth. Given such a picture, biogeographers ask the question 'What lives where and why?' The answers given invoke either ecological or historical explanations, or both.

Ecologists consider distribution patterns as communities at various levels of organization: species live in habitats belonging to ecosystems within biomes. For ecologists biogeography seems to be subsidiary to ecology. Indeed as Ball (1976) points out, some ecologists, such as MacArthur and Wilson (1967) do not see any distinction between the two subjects. We think no one would disagree that many similarities of climate, topography, and physiognomy exist between the rain-forests of South America and South-east Asia but when examined closely the organisms occurring in each rain-forest are usually quite different; species in similar niches belong to widely disparate taxa with unique histories that demand different explanations.

Some ecologists go so far as to say that the historical component is ecological (e.g. Stott 1981; Flenley 1979). This is because when looking at the dynamics of community structure—dispersion, on a local scale in a short time frame—changes in light, temperature, and humidity produce profound differences in the relative frequencies of occurrence of species. Ecological biogeography neglects evolutionary, major disjunctional, and long-term temporal components which together comprise the elements of historical biogeography. Historical biogeographers focus on older events—the development of biotas on a world scale—in an effort to study the history of the earth rather than the interactions of species in communities.

The data for historical biogeography come from comparative biologists and systematists who include distributional information in their monographs. A major synthesis is found in the work of Croizat, especially in *Space, time and form, the biological synthesis* (1964)—a book of some 800 pages regarded by the author as a mere summary of his earlier views. Croizat was a biologist who considered that geological events and the 'form-making' of species are part of one historical process. Despite Corner's (1959) perceptive observation that Croizat's work is critical for anyone interested in the idea that the world and its biota evolved together, only recently has notice been taken of his work. The interest has come at the same time as the gradual acceptance of plate tectonics and mobilist concepts of earth history and widening interest in the use of cladistics for phylogeny construction, a method outlined formally by Hennig (1950, 1965, 1966) and later developed by others. Recent expositions of the combination of the work of Croizat and Hennig can be found in the papers of Ball (1976), Nelson (1978*b*), Croizat *et al.* (1974), Rosen (1976, 1978, 1979), and Nelson and Platnick (1981). However, Croizat (1982) emphasized strongly that he never endorsed Hennig's work on either cladistics or biogeography. Neither did he endorse the integration of cladistics with panbiogeography. As such, the cladistic concept of historical biogeography is separate from that of contemporary panbiogeographers such as Craw (1982), who follow Croizat's method.

As Nelson (1978*b*) and Nelson and Platnick (1980) point out, biogeography 'is a peculiar discipline because most of its practitioners are not biogeographers but systematists specializing on some group of organisms'. In much the same way as systematists describe species and study their interrelationships on the basis of character distributions to discover what groups exist and what their origins might be, one could say that historical biogeographers study the distribution of species in different areas of the world to see how the areas are interrelated, what groups (regions) exist and what their origins might be (see Platnick and Nelson 1978).

Concepts of biogeographic regions, their relationship and general patterns have changed during the history of the subject. Ball (1976) and Patterson (1981*a*) recognize three phases of development in systematics (which Ball terms the descriptive phase, the narrative phase, and the analytical phase), all of which have influenced historical biogeography.

In fact, an analogy may be drawn between the history of taxonomic methods and biogeographic methods. Descriptive, or alpha taxonomy is the data-gathering enterprise of systematics which seeks to distinguish and describe species. Narrative taxonomy, by contrast, attempts to explain the evolutionary relationships of different organisms by using what may be called evolutionary systematic or omnispective methods for assessing groups by weighted similarities. Narrative taxonomy is common to many undergraduate textbooks where evolutionary relationships are presented as stories rather than as scientific hypotheses. Analytical taxonomy in this book refers explicitly to cladistic methods as originated by Hennig (1950, 1965, 1966) and developed somewhat independently by Wagner (1961, 1980), Camin and Sokal (1965), Kluge and Farris (1969), and Farris (1970). Modern textbooks on cladistics include that of Eldredge and Cracraft (1980) but more important for biogeography, Nelson and Platnick (1981) and Wiley (1981).

Historical biogeography has been associated mostly with the narrative and analytical phases of taxonomy and can similarly be divided into narrative biogeography and analytical biogeography. Narrative biogeography is the practice of using historical (geological or climatic) events and various *ad hoc* assumptions, such as that all taxa have a 'centre of origin' and dispersed to other areas, as a basis for explaining distribution patterns. Analytical biogeography, by contrast, refers to the comparison of the pattern of relationship of different groups of organisms occupying similar areas to find biogeographical patterns.

Many names have been applied to analytical historical biogeography which emphasize explicitly the phylogenies of plants and animals; these include vicariance biogeography, cladistic biogeography, phylogenetic biogeography, or even simply historical biogeography. We have chosen *cladistic biogeography* as the title of the method of historical biogeography we endorse. It combines cladistics with vicariance biogeography, and has as its basic premise the search for patterns of relationships among areas of endemism.

Our aim is to present a comprehensive review of the theory and method of cladistic biogeography, summarizing the historical changes leading to its formulation and presenting practical examples for the systematist who may wish to incorporate biogeography into taxonomic revisions.

Chapter 1 concentrates on areas and distribution, the concepts of regions and the relationships of regions in a brief historical résumé. The main steps in the refinement of method are outlined in Chapter 2. Chapter 3 deals with the application of new methods when applied to the real world and Chapter 4 looks at one of the intellectual challenges of biogeography—a rational explanation for the history of biota around the Pacific basin.

We prefer to call this book a monograph rather than a textbook. We have discussed and debated our ideas with colleagues who, in this case, are mostly practising biogeographers. However, we are presenting our own interpretation of a field that has grown with the input from many of these colleagues, whose work we cite herein. Cladistic biogeography remains a changing field; we have written this book so that discussion, debate, and practice of biogeography may extend to a wider audience, to all systematic biologists.

1 HISTORICAL BIOGEOGRAPHY

1.1 INTRODUCTION

The aim of this chapter is to review briefly the history of biogeography up to the late 1960s. The following three chapters detail the developments in cladistic biogeography since that time. The geographic distribution of living things has always intrigued biological theorists, especially systematists, since the fifteenth and sixteenth century voyages of exploration brought to light exotic plants and animals. As the acquisition of specimens and classifications improved, especially in the nineteenth century, botanists and zoologists tried to understand the burgeoning wealth of distributional evidence. As it became clearer that all organisms had particular distributions, enquiring minds tried to establish the historical causes of such patterns. The complexity of biogeographic patterns is made even more complex by the arguments of biogeographers. For example, the importance of ecological causes as compared to historical causes creates confusion. Differences in the alternative scientific approaches of dispersal and vicariance biogeographers obfuscate the aims of historical biogeography.

Cladistic biogeography is a method of historical biogeography that combines Hennig's (1950) notion of relationship with distribution patterns to ask the question 'Are areas of endemism interrelated among themselves in a way analogous to the interrelationships of the species of a certain group of organisms?' In other words, of three different areas definable by the endemic taxa occurring in those areas, are two of them more closely related to each other than the third? Such a question is part of a historical tradition of empirical enquiry which started in the work of Humboldt, de Candolle, and Croizat. It is an alternative to dispersal biogeography which has its roots in the work of Linnaeus, Darwin, Wallace, and Simpson. This chapter attempts to describe the differences between the two approaches.

1.2 DISTRIBUTION PATTERNS

Every species has a particular distribution. Species vary in range, in frequency of occurrence and degree of geographical continuity. For example, consider the natural distribution of the dandelion, *Taraxacum magellanicum* in the southern hemisphere. The map (Fig. 1.1) shows a fragmented distribution separated by oceans and the black areas represent the boundary limits of the species. The second map (Fig. 1.2) shows the sub-cosmopolitan distribution of the waterweed hornwort, *Ceratophyllum demersum*, occurring on all the major continents. Exact localities are shown only for a few areas, i.e. Northern Canada, the Caribbean, and Western Pacific, but the range is shown by the shaded black areas. There are conspicuous gaps on the map from which one can only conclude that it is an aquatic species absent from dry areas. The third example, is the distribution of the North American-South American species *Osmorhiza chilensis* of the carrot family (Umbelliferae) (Figs 1.3 and 1.4). The map is generalized showing the limits of distribution in Eastern North America and South America and scattered localities across temperate North America.

All three species survive and reproduce in accordance with their own environmental requirements. One does not find, nor expect to find, a freshwater angiosperm living with the arctic terrestrial dandelion, and neither could co-occur with the Umbellifer, *Osmorhiza*.

Each species occupies a precise area, and what is true for one species is also true for groups of species—genera and families. Consider the generalized distribution of *Acaena* (Rosaceae), a large genus of some 100 species of mostly herbs

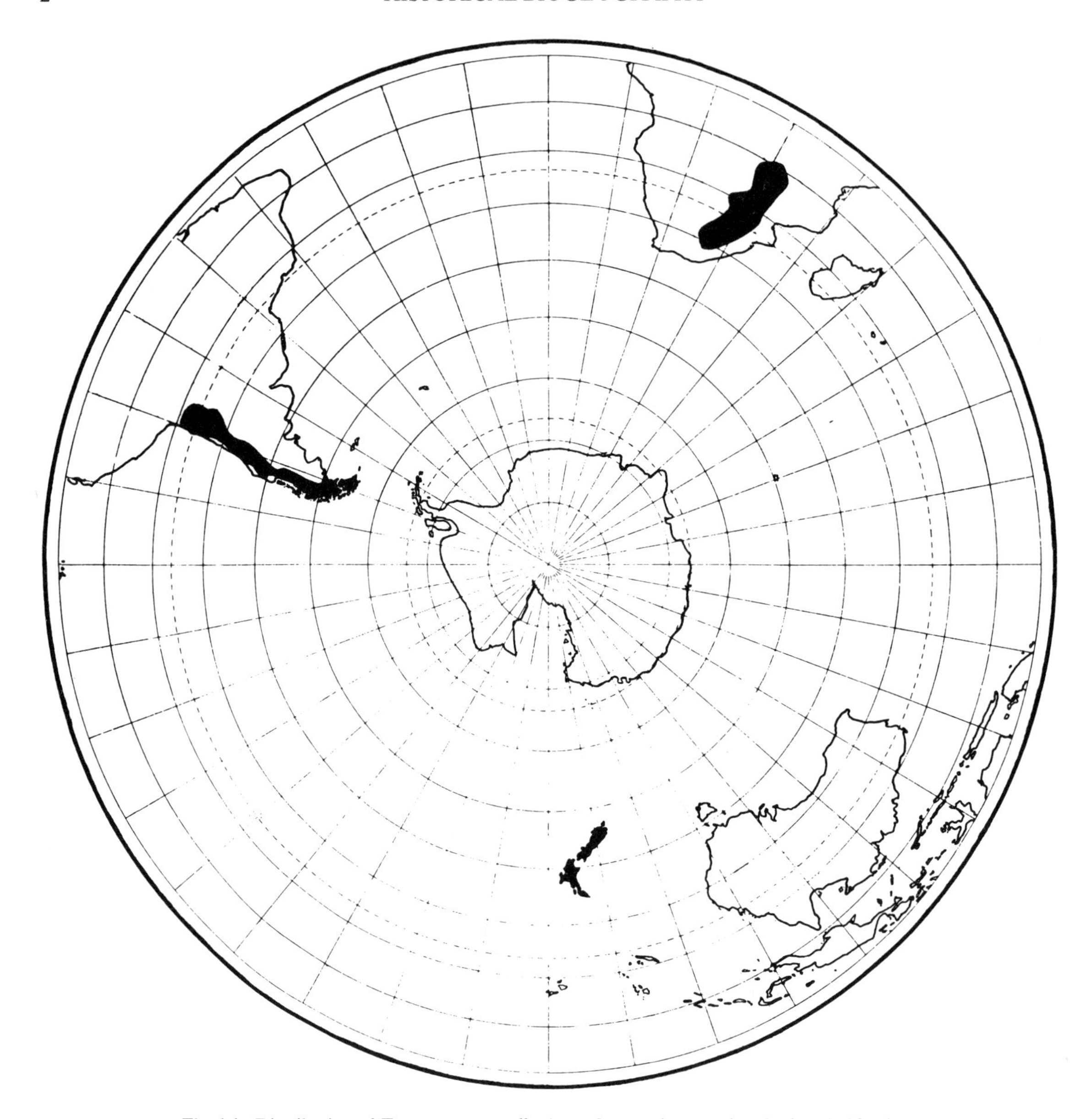

Fig. 1.1. Distribution of *Taraxacum magellanium* (Compositae). (After Croizat 1952, Fig. 9.)

occurring in South temperate areas, Antarctica, the Andes, Hawaii, and California (Fig. 1.5). Many species occur in much the same area as *Taraxacum magellanicum*. We may ask the question, does this mean that *Taraxacum magellanicum* and *Acaena* have a similar history to show this similar pattern? As a second example the map in Fig. 1.6 shows the generalized distribution of the pantropical group of freshwater fishes of the Gobiidae. This group occurs in much the same parts of the tropics and warm temperate areas as *Ceratophyllum* but is largely absent from northern and southern cool temperate areas.

Finally consider the distribution of the genus

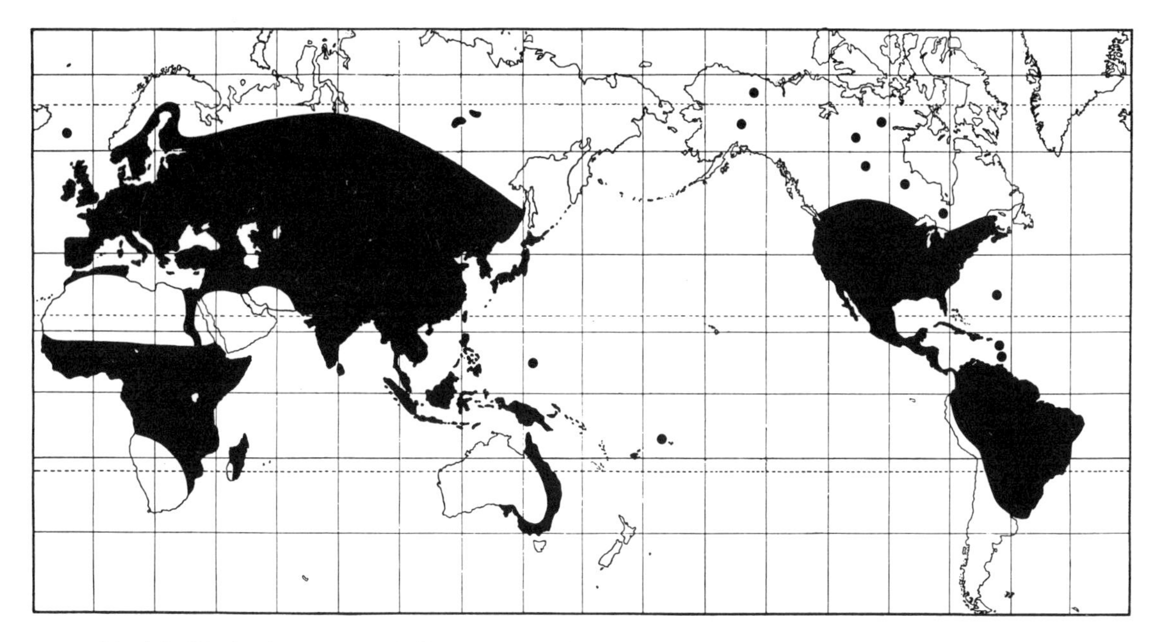

Fig. 1.2. Distribution of *Ceratophyllum demersum* (Ceratophyllaceae). (After Thorne, 1972, Fig. 43, p. 397.)

Empetrum (Fig. 1.7), with just two species occurring in a very disjunct bipolar distribution. The range of the southern crowberry, *E. rubrum* corresponds almost exactly with the southern populations of *Osmorhiza chilensis* (Fig. 1.3). The nothern crowberry, *E. nigrum*, has a range which partially overlaps that of *Osmorhiza* but is more widespread throughout North America, Eurasia, and even the Arctic.

Although not much can be said about the individual distributions of each taxonomic group except that also they have unique ecologies, alternate explanations are required to account for the sympatric occurrence of the widely different groups. For any taxon, details of its distribution patterns are usually acquired very slowly by ecologists and systematists, and even today we have often only the vaguest impressions of true distribution patterns. However, despite the vagaries of any one taxon we have sufficient evidence to show that some patterns are repeated by many unrelated groups. For example, the southern hemisphere temperate regions of South Africa, southern South America, Australia, Tasmania, and New Zealand, together with the islands of subtropical New Caledonia and tropical New Guinea, show many strikingly similar allopatric disjunctions in related taxa. The taxa of unrelated groups in any one region provide many examples of sympatric occurrence (see Table 1.1). In other words, one interesting aspect of individual distributions is that they define vicariant patterns.

Vicarious distribution patterns are more common among closely related taxa, such as species within a genus, or subspecies within a species, but are much less common between genera within a family or families within orders and so on. Groups at the family level or above tend to be sympatric, meaning they are greatly overlapping in the same area. For example, the southern beech genus *Nothofagus* (Fagaceae) is broadly sympatric with the Winteraceae family (see Table 1.1), a distantly related angiosperm family.

Disjunct vicarious distributions occurring repeatedly in different groups for the same areas are not confined to the southern hemisphere but

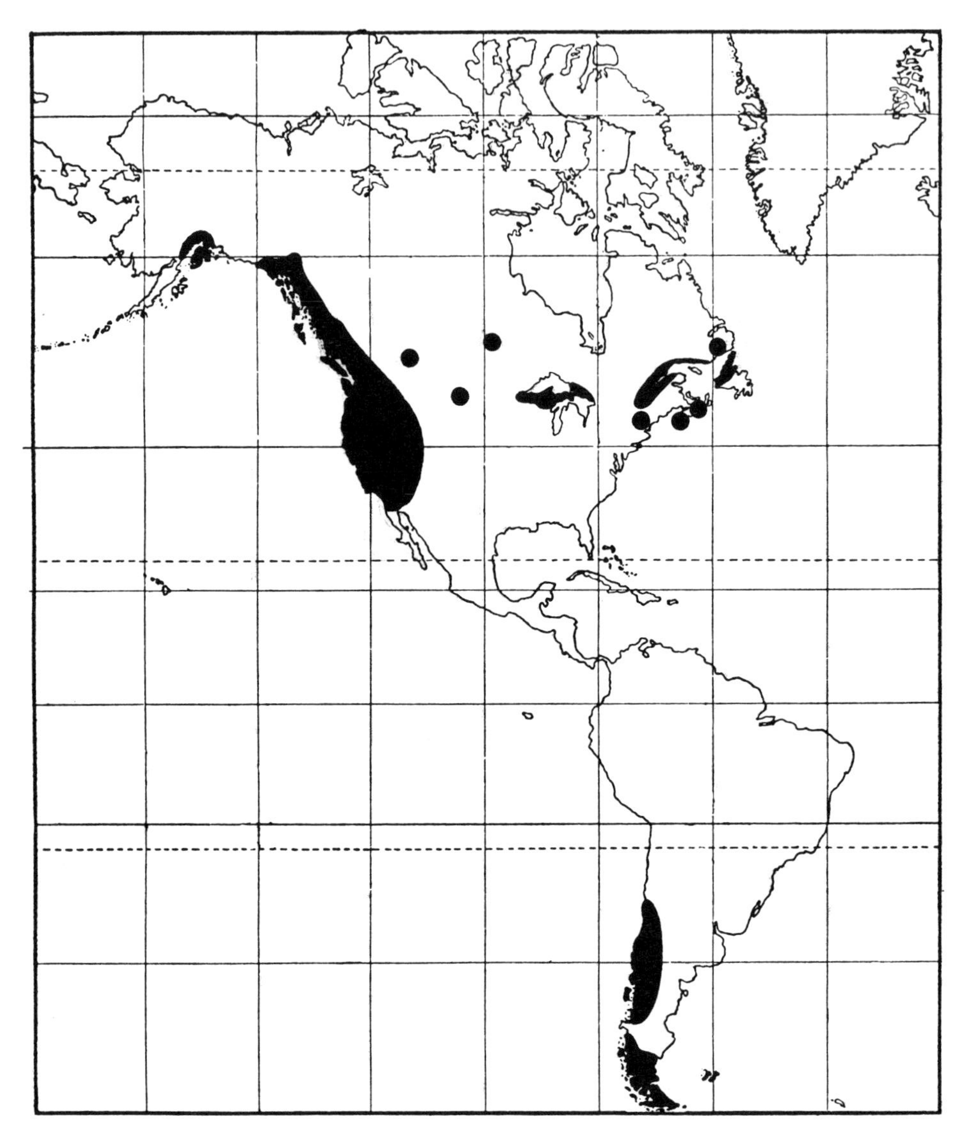

Fig. 1.3. Distribution of *Osmorhiza chilensis* (Araliaceae). (After Thorne 1972, Fig. 35, p. 393; see Fig. 1.4.)

exhibit a global pattern (Table 1.2). Classifications of intercontinental disjunctions for plants can be found in Croizat (1952), Good (1964, 1974) and Thorne (1972).

Individual distributions, although in some cases widespread, usually have narrow ranges when viewed on a map of the world. More importantly, many distributions of particular taxa fall into a pattern that can be called endemic, especially apparent when, say, a species is confined to one continent, a small oceanic island or a mountain top.

The sympatric distributions of unrelated taxa leads to a concept which we can call 'areas of endemism'. Thus, as examples, we can talk about the Macaronesian element of the Canary Islands,

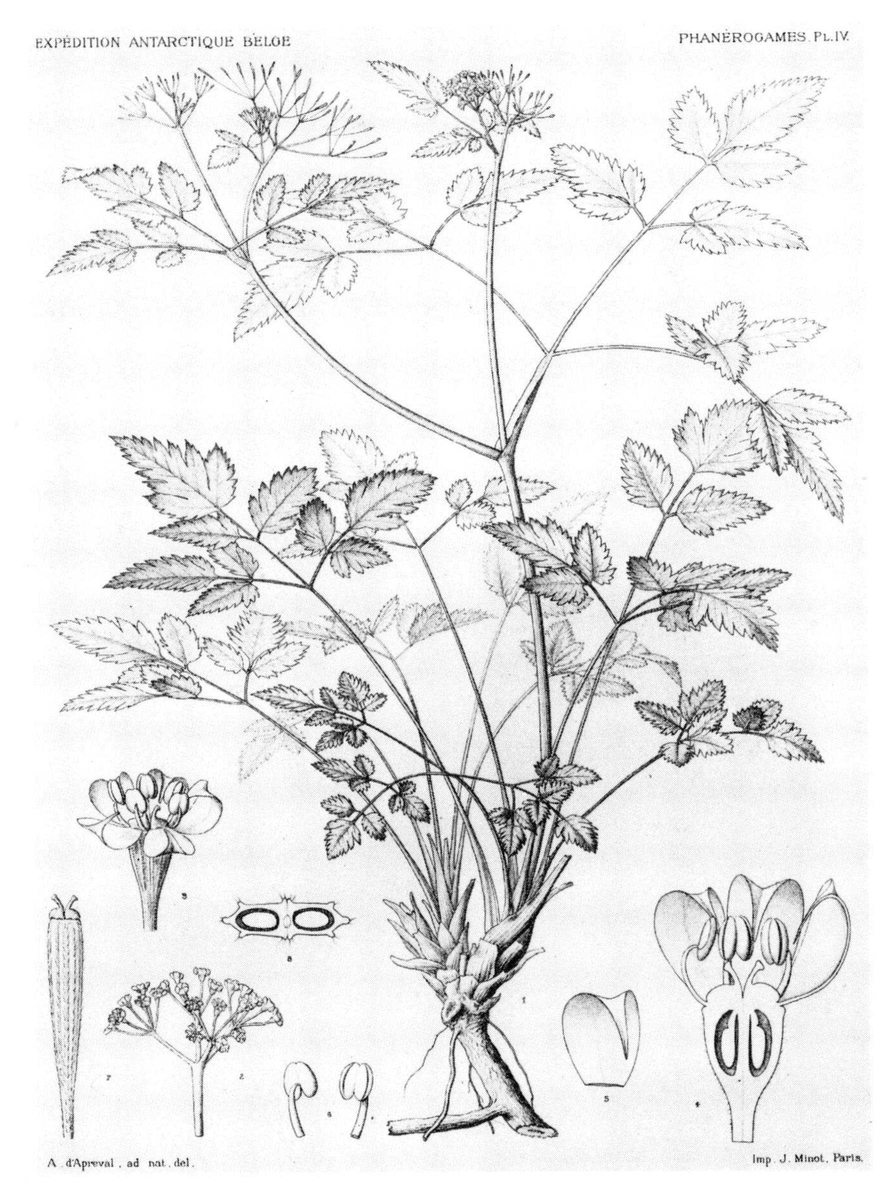

Fig. 1.4. *Osmorhiza chilensis* (Hook and Arn). A member of the carrot family (Umbelliferae) with a bipolar distribution in North and South America (see Fig. 1.3; de Wildeman 1905, Plate iv).

the marsupials of Australia, the insects of New Guinea, and so on.

What do areas of endemism mean? An examination of the ecological conditions of each species in one area of endemism does not really tell us anything about sympatry. For example, why do southern beech trees show broad distributions similar to chironomid midges when neither is necessarily dependent on the other? Is it that both have independently become adapted to similar

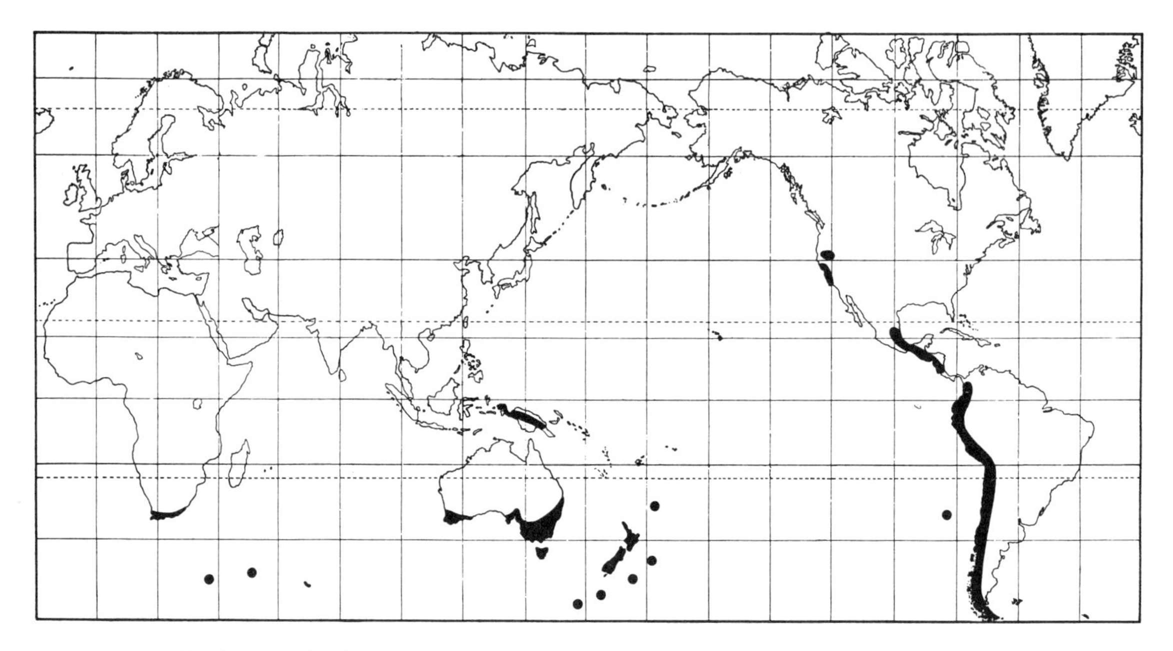

Fig. 1.5. Distribution of the genus *Acaena* (Rosaceae). (After Thorne 1972, Fig. 41, p. 396.)

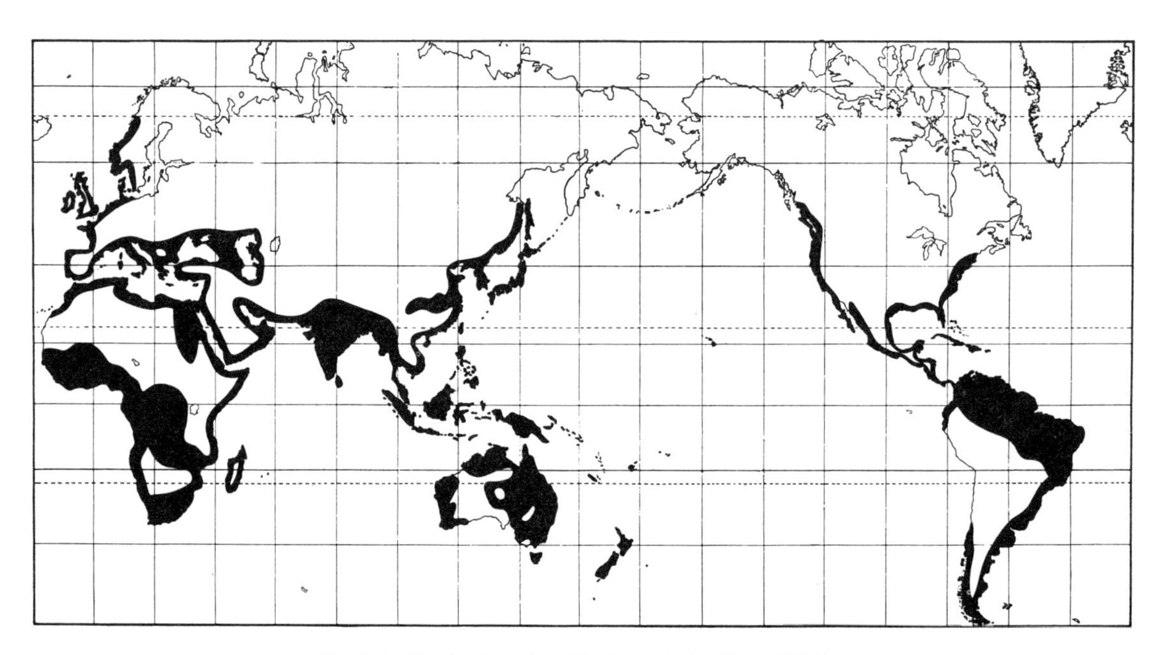

Fig. 1.6. Distribution of the Gobiidae. (After Berra 1981.)

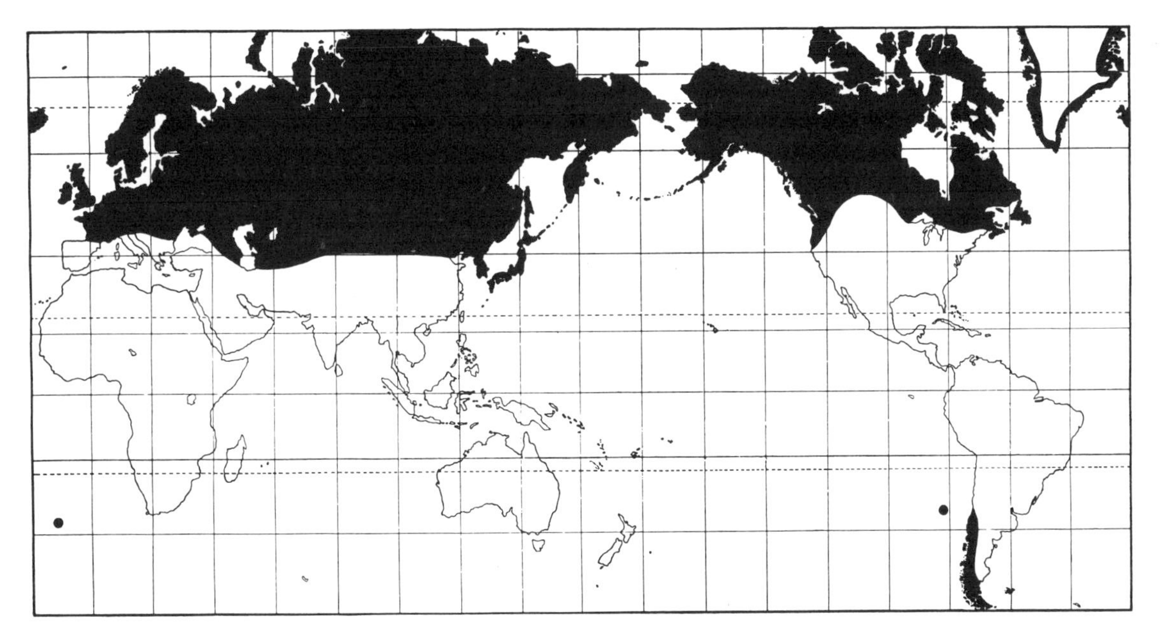

Fig. 1.7. Distribution of the genus *Empetrum* (Empetraceae). (After Thorne 1972, Fig. 36, p. 393.)

southern hemisphere habitats or does the history of the biotas in which they occur provide a different answer?

1.3 A HISTORY OF IDEAS

1.3.1 Origins

That every organism has a particular distribution pattern has concerned biologists for two centuries. Modern systematics is generally regarded to have started with the work of Carolus Linnaeus (1707–78), a major contributor to the theory of hierarchical classification, and one of the earliest writers to interpret the distribution of organisms. Linnaeus believed that the earth was changing, with the continents growing in size and concluded that if this process had operated in the past:

> the continent in the first ages of the world lay immersed under the sea, except a single island in the midst of this immense ocean; where all animals lived commodiously, and all vegetables were produced in the greatest luxuriance. (Translated from Linnaeus 1781, p. 77, after Nelson and Platnick 1980.)

He realized that organisms had various ecological requirements and suggested that the primordial island must have been in the tropics and 'bore a very lofty mountain' inhabited at different altitudes by species with different ecologies. As new land emerged when the seas receded, organisms migrated by various means to colonize those parts with suitably similar ecological conditions (Nelson and Platnick 1980). Within these comments were two important ideas: that species originate and disperse from a 'centre of origin', and that regularities in distribution were controlled by ecological conditions. A development and better expression of these ideas appears in Willdenow's *Grundriss der Kräuterkunde* (1798), part seven:

> By history of plants is meant the influence of climate on vegetation, the changes which plants have probably undergone as a result of the revolutions which have taken place on our globe, their distribution over the earth's surface, their migrations, and, lastly, the provisions nature has made for their preservation. (Translated from Wulff 1950.)

Table 1.1

Distribution of endemic taxa in South America (SA), Africa (Af), Madagascar (M), Tasmania (Tas), Australia (Aus), New Zealand (NZ), New Guinea (NG), and New Caledonia (NC)

	Areas								
Family/genus	*Af*	*M*	*SA*	*Aus*	*Tas*	*NZ*	*NG*	*NC*	*Others*
Chironomid midges	+		+	+		+			1, 2
Winteraceae		+	+	+	+		+	+	3, 5
Coriaria			+			+	+	+	1, 2, 3, 4
Proteaceae (Gevuina; Lomatia; Oreocallis and Orites combined)			+	+	+		+		
Acaena	+		+	+	+	+	+	+	3
Osteoglosine fishes			+	+			+		
Ratite birds			+	+			+		2
Stylidiaceae			+	+	+	+			
Nicotiana			+	+				+	1
Hylid frogs and Chaleosyrphus (Syrphid flies)			+	+			+		1, 2
Marsupials (Recent)			+	+			+		1
Nothofagus			+	+	+	+	+	+	

Other areas; 1, N. America; 2, Europe; 3, Central America; 4, China/Japan; 5, Malaysia.

According to Wulff (1950), a number of questions were in Willdenow's mind—did the seas we see today formerly occupy much greater areas? Indeed was not the earth once almost entirely covered with water, from which projected mountain peaks, then the only habitats available for terrestrial organisms? As the seas receded land area increased, and plants and animals dispersed from the initial habitats. Later, great geological climatical events such as hurricanes, earthquakes, and volcanoes destroyed life over huge areas. This is evidenced especially by the distribution of related endemics in widely separated areas.

Lands now separated by oceans may, in former epochs, have been united. ... Thus, the northern part of America may have been connected with Europe, New Netherlands (Australia) with the foothills of the Cape of Good Hope. (Willdenow after Wulff 1950.)

This is a prophetic passage with the idea that in addition to migrations present day disjunctions might be caused by the separation of formerly continuous biotas. Despite such a concise statement, it is generally regarded (e.g. Wulff 1950; Nelson 1978*b*) that Alexander von Humboldt is the founder of phytogeography and Compte de Buffon the founder of zoogeography. De Buffon (1776) had already formulated the general principle that animal life was very different between different areas of the world, especially the mammal faunas of Africa and South America, when von Humboldt and Bonpland (1805) expressed some remarkable ideas about the history of the earth:

In order to come to a decision as to the existence in ancient times of a connection between neighbouring continents, geology bases itself on the analogous structure of coastlines, on the similarity of animals inhabiting them and on ocean surroundings. Plant geography furnishes most important material for this kind of research. It can, up to a certain point, determine the islands which, at one time united, have become separated from one another; it finds the separation of Africa and South America occurred before the development of living organisms. It is again this science that shows which plants are common to both eastern Asia and the coastlands of Mexico and California, and whether there are some which grow in all zones and at all altitudes. It is by the aid of plant geography that we can go back with some certainty to the initial physical state of the globe. It is this science which can decide whether, after the recession of the waters to whose abundance and movements the calcareous rocks attest, the entire surface of the earth was covered simultaneously with diverse plants, or whether according to the ancient myths of various peoples, the globe, having regained its repose, first produced plants only in a single region from which the sea currents carried them progressively, during the course of centuries, into the more distant zones. (von Humboldt and Bonpland 1805, pp. 19-20, after Wulff 1950, p. 11.)

Humboldt and Bonpland were convinced that the history of organisms and the history of the earth go together and the Linnaean idea that organisms originated in one area and migrated to others was ill-founded. Besides distribution as a key to the past, Humboldt and Bonpland also suggested that the examination of fossils would give not only clues about the migration of plants over the globe but also some clues as to what past climates would have been like:

In order to solve the great problem as to the migration of plants, plant geography descends into the bowels of the earth: there it consults the ancient monuments which nature has left in the form of petrifications in the fossil wood and coal beds which constitute the burial places of the first vegetation of our planet. (von Humboldt and Bonpland 1805, p. 22 after Wulff 1950, p. 11.)

Thus, by the turn of the 19th century, three sets of ideas existed:

(a) that organisms could originate from one (or more) 'centres of origin' and migrate to other parts of the globe;

(b) that changes in the world itself—earth history—could explain differing distribution patterns; and

(c) that different habitats and climates could determine the perceived present day pattern. The principal point to emerge was the generality of de Buffon's original observation, that the fact that different areas, especially the new and old world tropics, contain different species is true for all organisms and not just for mammals.

Table 1.2
A broad classification of major distribution patterns of seed plants (after Thorne 1972; Stott 1982)

I	Eurasian-North American
	1 Arctic
	1a Circum-Arctic
	1b Beringian-Arctic
	1c Amphi-Atlantic-Arctic
	2 Boreal
	2a Circum-boreal
	2b Beringian-boreal
	2c Amphi-Atlantic-boreal
	3 Temperate
	3a Circum-north temperate
	3b North-south temperate
	3c Fragmentary-north temperate
II	Amphi-Pacific tropical
III	Pantropical
IV	African-Eurasian (-Pacific)
	1 African-Mediterranean
	2 African-Eurasian
	3 African-Eurasian-Malesian
	4 African-Eurasian-Pacific
	5 African-Eurasian-Australasian
	6 Indian Ocean-Eurasian
V	Amphi-Indian Ocean
VI	Asian-Pacific
	1 Asian-Papuan
	2 Asian-Papuan-Melanesian
	3 Asian-Papuan-Pacific Basin
	4 Asian-Papuan-Australasian
VII	Pacific Ocean
VIII	Pacific-Indian-Atlantic Ocean
IX	American-African
X	North American-South American
XI	South American-Australasian
XII	Temperate South American-Asian
XIII	Circum-south temperate
XIV	Circum-Antarctic

1.3.2 Habitations

Together with the acquisition of additional data on the distributions of particular organisms came advances in the classification of distributions as evidenced in the work of A. de Candolle. He distinguished between two branches of plant geography—the equivalents of modern ecological and historical biogeography. De Candolle (1820) discussed the influence of external elements, temperature, light, and humidity in determining stations, or habitats as we call them today, and also different types of plant distribution in 'habitations' or areas. Ecological biogeography is the study of stations; historical biogeography the study of habitations. De Candolle thought confusion between the two aspects would undermine the development of either science. It was clear to de Candolle that the factors which determined stations were quite different from those which determined habitations. He also noted that cosmopolitan species were exceptions to the general law of distribution—namely that different species occur in different habitations. He concluded that cosmopolitan species transcended habitation boundaries by dispersal and noted that plant seeds could be transported by the continual action of three causes—water, wind, and animals (including humans). This could explain why certain species appear to occur in different continents. To account for particular distribution patterns de Candolle deduced that all plants originated in a particular region.

His classification is shown in Table 1.3. The list is a good approximation of some areas of endemism that we recognize today.

Of interest to us is the fact that de Candolle's comparative studies of the regions allowed him to recognize endemic genera with many species occurring in one area and that certain genera occur in widely disjunct regions. As one explanation of disjunction and endemism, de Candolle resurrected Willdenow's idea that mountains form barriers between regions and rejected the idea that they could be a 'centre of origin'. He did not suggest that mountains and continental break-up could cause the divisions between regions, because if that were the case one would expect the same species on either side of a mountain range. Instead, he criticized the immutability of species, inherent in theories of geographical botany at that time, and considered the origin of species to be due to external forces as yet unknown.

The important point to emerge from de Candolle's work is that his regions were roughly equivalent to what are recognized today as areas of endemism, and a precise statement of de Buffon's original observation. Subsequent studies in historical biogeography took two quite different courses. On one hand, as evidenced particularly in the work of Forbes, there was an interest in considering the history of earth and its biota together, whilst on the other, as in the work of Darwin, there was a separation of the two.

Table 1.3
de Candolle's twenty botanical regions

1	Boreal Asia, Europe, and America
2	Europe south of the boreal region and north of the Mediterranean
3	Siberia
4	The Mediterranean region
5	Eastern Europe to the Black and Caspian Seas
6	India
7	China, Indo-China, and Japan
8	Australia
9	South Africa
10	East Africa
11	Tropical west Africa
12	Canary Islands
13	Northern United States
14	North-west coast of North America
15	The Antilles
16	Mexico
17	Tropical America
18	Chile
19	Southern Brazil and Argentina
20	Tierra del Fuego

1.3.3 One history

As a starting point for his exposition on the history of the British fauna and flora, Forbes (1846) assumed the existence of 'specific centres' or 'certain geographical points from which the individuals of each species, originating from a single progenitor or two, began their geographical distribution.' As an explanation for particular distributions, duckweed in ponds, alpine herbs in mountains, and trees in forests, Forbes concluded that organisms are adapted to particular conditions and will migrate only into those places of which they are capable. Examining the interrelationships of the British flora, the various different elements, according to Forbes, could have only come about by the migration of species from continental areas prior to the separation of the British Isles from the mainland. Of interest to us is that Forbes arrived at his conclusions using the evidence that several similar or identical species in disjunct areas are best explained not by repeated dispersals but by one isolation event.

These ideas were developed further by the botanist J. D. Hooker to provide general explanations for the phytogeography of the whole earth—a transition from the study of small, separate units to entire continents. Hooker's biogeographical studies ranged from the Galapagos to the floras of the southern hemisphere and the tropics. Hooker interpreted the distribution of organisms in a most plausible manner. Thus, the origin of organisms on oceanic islands, such as the Galapagos flora, could be explained as the results of the transport of its component species, especially the non-endemics, by transport on ocean currents, by wind, birds or humans. The individual appearances of island taxa were due to later modifications under the influence of isolation. However, his best known contribution came later in his introductory essays to his 'Botany of the Antarctic voyage'. In volume II, the Flora of New Zealand (Hooker 1853), by using Lyell's principle that each species can only have arisen at one point on the globe, Hooker reasoned that the widely disjunct islands of the Antarctic, with very similar floras, must have formed a single area—a land mass occupying a continent larger than that in the Antarctic ocean. In Volume III of the work (Hooker 1860), devoted to the flora of Australia and Tasmania are two more sections of interest to us: 3, 'On the general phenomena of distribution of plants in area' and 4, 'On the general phenomena of the distribution of plants in time'. In the former, Hooker writes that the present distributions of plants can only be explained by past events and to study them it is necessary to study past changes in climates and the distribution of dry land. In the latter he reviews palaeontological data to advance the proposition that changes in the earth's surface—lands replaced by sea and valleys replaced by mountains—take place in a very short time relative to the ages of biological groups.

1.3.4 Separate histories

By the middle of the 19th century it was becoming clearer that the earth and its biota changed with time and that a general theory of mutability of species was lacking. Darwin's *Origin of species* appeared in 1859, but surprisingly, he separated his mechanism of evolution from biogeography. In fact, only Chapters 12 and 13 are devoted to biogeographical questions yet they did greatly influence later ideas on distribution. Darwin opens Chapter 12 with the statement:

> In considering the distribution of organic beings over the face of the globe, the first great fact which strikes us is that neither the similarity nor the dissimilarity of the inhabitants of various regions can be wholly accounted for by climatal and other physical conditions. (Darwin 1859: p. 493.)

Darwin responded to de Candolle's 'Origin of species' problem by saying that the similarities of the biotas in southern hemisphere lands and the similarities of the North and South American organisms were not due to changes in the earth's surface but due to inheritance:

> We see in these facts some deep organic bond, throughout space and time, over the same area of land and water, independently of physical conditions. ... The bond is simply inheritance ... (Darwin 1859.)

The difference between species, the unique characters of varieties and different species according to Darwin, originate because each species is produced in one area, subsequently migrating from that area and changing by natural selection through time.

Chapter 13 is concluded with another profound statement:

> The endurance of each species and group of species is continuous in time; ... so in space, it certainly is the general rule that the area inhabited by a single species or by a group of species is continuous and the exceptions, which are not rare ... be accounted for by former migrations under different circumstances, or through occasional means of transport, or by the species having become extinct in the intermediate tracts. (Darwin 1859, p. 564.)

Thus, Darwin proposed a method of biogeography that could account for all possible distributions. The finding of similar species in the British Isles and in Europe was easy to comprehend, since until relatively recent time they formed a continuous land surface. At the same time, de Buffon's difference in biotas between Africa and South America, despite similar habitats is understandable, because they have been separated for a long time. The occasional existence of identical species separated by vast distances attests to great dispersal abilities. Darwin criticized Lyell and Forbes too when he wrote:

> Other authors have thus hypothetically bridged over every ocean and united almost every island with some mainland. If indeed the arguments used by Forbes are to be trusted, it must be admitted that scarcely a single island exists which has not recently been united to some continent. This view cuts the Gordian knot of the dispersal of the same species to the most distant points, and removes many a difficulty; but to the best of my judgement we are not authorised in admitting such enormous geographical changes within the period of existing species. (Darwin 1859, p. 505.)

In fact, Darwin allowed only certain types of earthly revolutions: 'great oscillations in the level of land and sea' and also 'existence of many islands, now buried beneath the sea, which may have served as halting places for plants and for many animals during their migration'. (Darwin 1859, pp. 505-6.) Thus, on the age of taxa Darwin had an opinion opposite to that of Hooker. Taxonomic groups are younger than the places they inhabit. To prove the point that dispersal must be the means for migration to oceanic islands, Darwin undertook a great many experiments on seed survival in sea water. The successful survivors he took as experimental proof of dispersal. Darwin thought that millions of years of chance dispersals, especially to those islands newly forming from the sea bed, would be sufficient to stock them with raw materials for future evolution. In other words, dispersal created areas of endemism. Darwin differed from de Candolle, Lyell, and Forbes, for whom dispersal accounted only for widespread or cosmopolitan species (Nelson 1978*b*).

With respect to the relative importance of dispersal, as distinct from the earth's physical features, to explain biogeography, Alfred Russel Wallace followed Darwin and focused on dispersal. His investigations were significant because the complex distribution of organisms over the earth was the outcome of both biological and 'physical' forces. To quote Wallace:

> The biological causes are mainly of two kinds—firstly, the constant tendency of all organisms to increase in numbers and to occupy a wider area, and their various powers of dispersion and migration through which when unchecked, they are enabled to spread widely over the globe; and secondly, those laws of evolution and extinction which determine the manner in which groups of organisms arise and grow, reach their maximum, and then dwindle away, often breaking up into separate portions which may survive in remote regions. (Wallace 1880, pp. 531-2.)

Wallace also subscribed to geological forces which could isolate whole biotas and changes in climate which were the main causes of extinction.

The issue for both Darwin and Wallace, and indeed for the subsequent Darwin-Wallace tradition so dominant for the last 100 years, was that because of dispersal, every taxonomic group had its own distributional history. Thus, for an area containing endemic species with nearest relatives elsewhere a special kind of 'isolating dispersal' is required to explain endemics. Such dispersal, once having been successfully achieved by a species, results in its isolation and subsequent differentiation into an endemic (Nelson 1978*b*). For Wallace there were two kinds of isolating dispersal, (a) dispersal over a pre-existing barrier and then isolation, and (b) dispersal of one species over a wide area followed by extinction of the intermediate populations to separate populations once forming a continuum. Despite philosophical shifts in earth history, systematics, and biogeography, the Darwin-Wallace tradition has continued more or less unabated into this century (e.g. Raven and Axelrod 1972, 1974). As Nelson (1978*b*) put it, the concentration on a class of improbable dispersals as an explanation for different taxa occupying different areas of endemism results in the 'science of the rare, the mysterious and the miraculous'.

1.3.5 Sclater's regions

The missing element from nineteenth century biogeography was the notion of the interrelationships of the areas in which individual groups of organisms occur. Sclater (1858) proposed a classification of the world based on the distributions of birds. Wallace (1876) considered that these regions, which he called 'Realms', might apply not only to birds but to animals in general and they have been used ever since (Fig. 1.8). For Wallace realms expressed the similarities of animal taxa in various areas of the world. Thus, an animal in Britain would generally be more like another animal from somewhere in the Palaearctic realm than from, say, the Ethiopian realm. These realms, although rather an abstract conception, did give some measure of the areas of the world that have each had a long and independent biotic evolution with overall taxonomic differences between the biotas of the different areas (Ross 1974). In other words, a vague assessment of the degree of relatedness of patterns on a global scale.

Most interestingly, Sclater realized the problem of relationship when he said:

> Little or no attention is given to the fact that two or more of these geographical divisions have much closer relations to each other than to any third. (Sclater 1858, p. 131.)

Sclater was not particularly satisfied with his six 'realms' but suggested that more systematic work would eventually allow us to arrive at the correct primary divisions of the globe. Curiously enough, despite here being all the ingredients for a method to find the generalities of biogeographical patterns, it remained 100 years or more before the

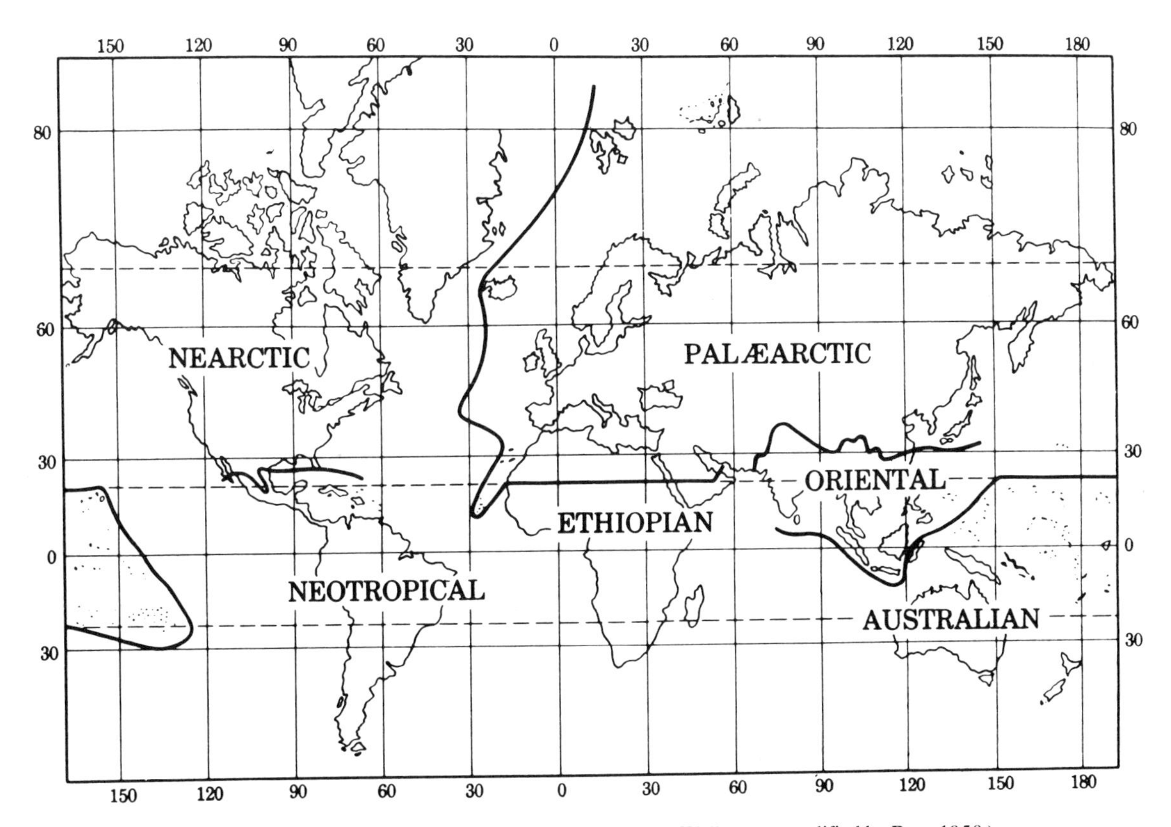

Fig. 1.8. Faunal regions (or realms). (After Sclater and Wallace, as modified by Ross 1950.)

necessary method could be realized. By the end of the eighteenth century it was known that species occur in different areas, that areas apply to both animals and plants; that there are biogeographical regions or areas of endemism and that they are related to one another. But much of this has been put to one side, and the main question for biogeographers in the twentieth century has been to determine the 'centre of origin' for different taxa, by applying evolution and dispersal as causal agents for distribution.

1.4 CENTRES OF ORIGIN

When considering the history of a biota in a particular area, the dispersal biogeographer asks: where did the group originate? and when and how did it come to occupy the place in which it lives today? For each question there are many possible answers and the choice depends entirely on what the author interprets as evidence. In phytogeography some of the most coherent attempts to identify centres of origin were made by Adams (1902), Willis (1922), Wulff (1950) and Cain (1944). Adams wrote that such criteria should be regarded solely as

> convenient classes of evidence to which we may turn for suggestions and proof as to the origin and dispersal of organisms . . . It should be clearly emphasised that it is the convergence of evidence from many criteria which must be the final test in the determination of origins . . .

Various assumptions have been used as guides to determine the 'centre of origin'; it may be the area in which the group of organisms displays the greatest diversity, or number of species, the area in which the most primitive forms occur, the area occupied by the most phylogenetically advanced or primitive members or where the oldest fossils occur. Cain listed thirteen criteria for recognition of centres of origin (Table 1.4). Each criterion may give a markedly different result. For example, Humphries (1981) discussed eleven different theories for the origin of the southern beeches, *Nothofagus* (Fagaceae), involving almost every continent; the favoured choices for a centre of origin have been North America, Europe, southeast Asia, New Caledonia, somewhere between Yunnan and Queensland, Antarctica, and New Zealand!

Table 1.4
Cain's (1944) criteria for determining 'centres of origin'

(a) The location of the greatest variety of forms of the taxon.
(b) The location of the area of greatest dominance and density of distribution.
(c) The location of the most primitive forms.
(d) The location of the area exhibiting the maximum physical development of individuals.
(e) The location of the area of maximum ecological productivity of the taxon.
(f) Continuity and convergence in the lines of dispersal.
(g) The location of least dependence on a restricted habitat.
(h) The identification of continuity and directness of individual variations or modifications radiating from a 'centre of origin' along the highways of dispersal.
(i) The area of origin indicated by natural geographical affinities.
(j) The direction of origin indicated by the annual migration routes of animals, especially birds.
(k) The region of origin indicated by seasonal appearance or general *phenology*.
(l) An increase in the number of dominant genes towards the 'centre of origin'.
(m) The concentricity of progressive equiformal areas.

As Forey (1981) noted, a theory about the 'centre of origin' may determine or be dependent upon a theory of a presumed dispersal route. The route from the centre to the present day whereabouts is usually made by reference to a variety of influencing facts: the occurrence of fossils, the age of the fossils, the past climate of the earth and so on. For example, to account for the resemblances in the North American and Asian palaearctic mammal faunas, Simpson (1962) suggested that there were four major exchanges between North America and Eurasia. The Bering Strait was the favoured highway and because of the cold climatic regimes at the presumed time of migration, only temperate rather than tropical mammals were to have moved. Besides fossils, so-called living fossils have also been used as evidence for origins of dispersal routes. Takhtajan (1969) asked the question: 'Can we in fact find out what part of the world was the cradle of the flowering plants?' His answer was, 'It was evidently a region in which they experienced a long period of evolution during which the principal families and many genera were differentiated, and it may also have been their centre of origin; in any case it was probably not very far from their birthplace'.

Some authors seek a 'cradle' in high latitudes with an arctic or antarctic origin (Heer 1868; de Saporta 1877) whilst others seek in the lower latitudes of the tropics or subtropics (e.g. Diels 1908; Kozo-Poljanski 1922; Wulff 1950; Smith 1970; Thorne 1972; Stebbins 1974). Bailey (1949) offered some advice: 'Look west, young man, towards the remnant of Gondwanaland'. According to Takhtajan (1969), the striking restriction of 'primitive' angiosperms to the islands and borders of the Pacific Ocean indicates that 'It is here, in eastern and south-eastern Asia, Australasia and Melanesia that the cradle of the angiosperms must be sought'. For Takhtajan the conclusion rests on the occurrence and preponderance of primitive forms; that is, those plants believed to possess primitive characters. Thus, he believes it was largely from the western Pacific that colonization of the world by angiosperms has taken place since the Cretaceous because the 'oldest' angiosperm fossils are to be found in the Barremian of the Cretaceous (see Hughes, Drewry, and Laing 1979) and the greatest number of 'living fossils' reside around the Pacific.

A third, but less specific place of origin is one with a mesophytic climate, as suggested by Stebbins:

> The most progressive vascular plants of the Jurassic period were probably the angiosperms which, if they existed today would be classified as belonging to or related to the Magnoliales. A logical assumption, therefore, is that the pioneer, ecotonal and mosaic habitats found in semiarid, sub-tropical mountainous regions were at that time occupied by the original ancestral Magnoliales and Dilleniales. (Stebbins 1974, p. 205.)

Raven and Axelrod (1974, p. 635) are a little more precise as to the exact location of such a place and suggest that west Gondwanaland might be the 'primary area of evolution' for 'many orders of angiosperms, and perhaps the earliest angiosperms themselves.'

By what means the organisms migrate from the 'centre of origin' is a newer question and is related to the identification of hurdles which inhibit movement. Simpson (1953) recognized three types of barriers to migration, recognized by degree of faunal similarity between areas:

(i) corridors—obvious heavy load migration routes—are indicated by high similarity;

(ii) filters—for example, the big deserts, water gaps and narrow strips of land, indicated by low similarity allowing only light load traverses; and

(iii) 'sweepstake' routes—the truly formidable barriers such as major oceans—which only allow the rare chance crossings.

Sweepstake routes are usually invoked to explain 'unbalanced biotas'—depauperate flora and fauna showing peculiar or ecologically unbalanced admixtures of organisms.

The principles of dispersal biogeography have remained consistent over the last century. Because dispersal hypotheses reside in a narrative framework they are irrefutable. Migrations have simply been modified to fit in with new views of earth history (Peake 1982). Dispersal hypotheses do not attempt to provide a general theory of earth history but rather individual case histories for each taxonomic group. Consequently, dispersal biogeography is a discipline divorced from earth history and invariably couched in the possible processes which could give rise to modern distribution patterns. Dispersal biogeography is an *ad hoc* discipline since it always requires external causes to explain the patterns (Croizat *et al*. 1974; Platnick and Nelson 1978; Patterson 1981*a*; Nelson 1982; Nelson and Platnick 1981). Consequently, dispersal hypotheses can never let us discover the history of the earth.

1.5 CROIZAT'S TRACKS

Although much twentieth century biogeography was, and indeed still is, clearly in a Darwin–

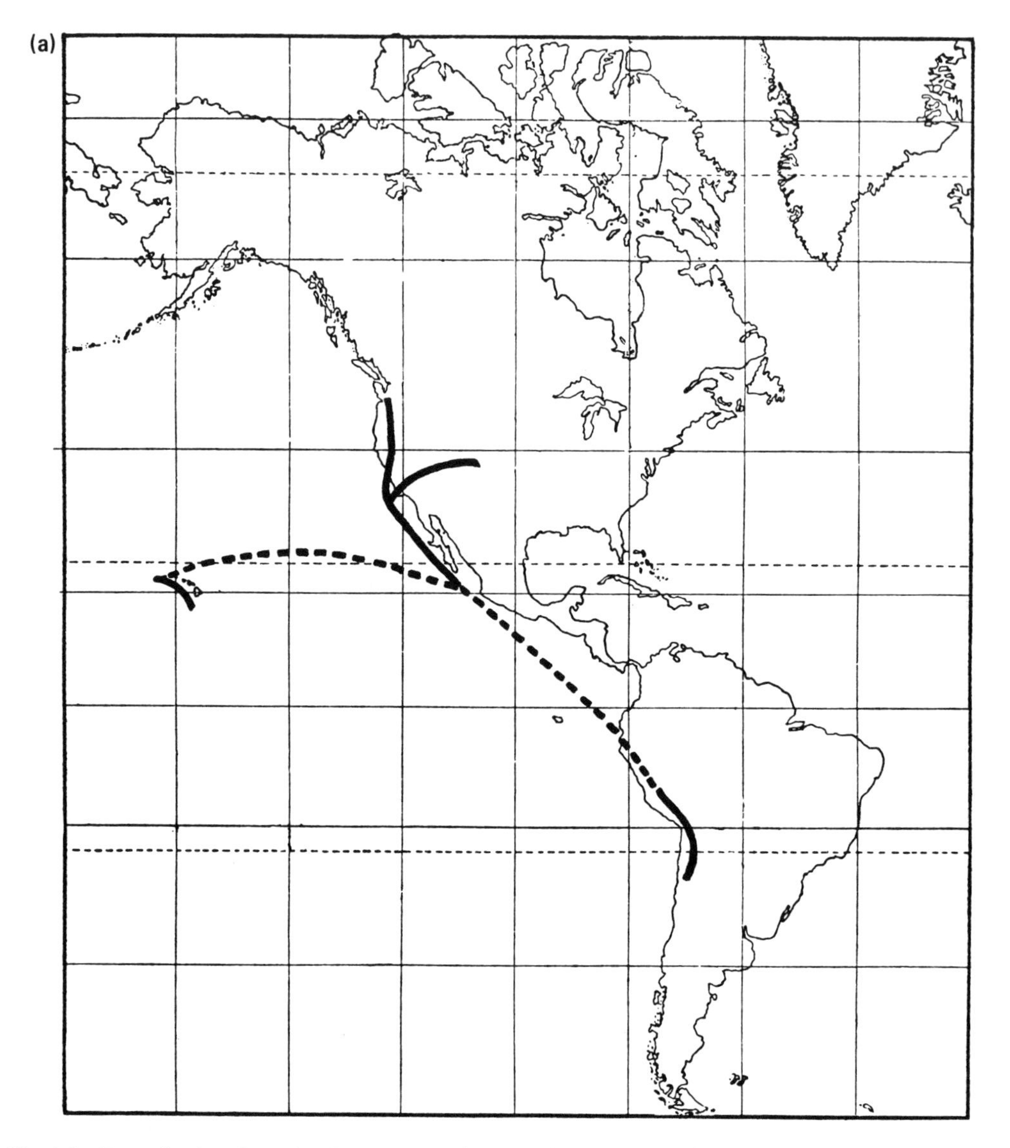

Fig. 1.9. Generalized tracks. (After Croizat 1964, Figs 1 and 2). (a) Geographic distribution. (b) Generalized track.

Wallace tradition as best shown by the writings of Matthew (1915), Simpson (1953, 1965), Darlington (1957, 1965), and Cain (1944), from time to time alternatives to dispersal have appeared; e.g. Wulff (1950), Croizat (1952, 1964), Melville (1981), and Craw (1982). Croizat (1952, 1958, 1964) undertook a laborious task in examining Sclater's problem—which areas of endemism are more closely related to one another in terms of other areas—through a method which he called panbiogeography—the biogeography of all organisms throughout the world.

Croizat's view of earth history and evolution may be reduced to two general principles:

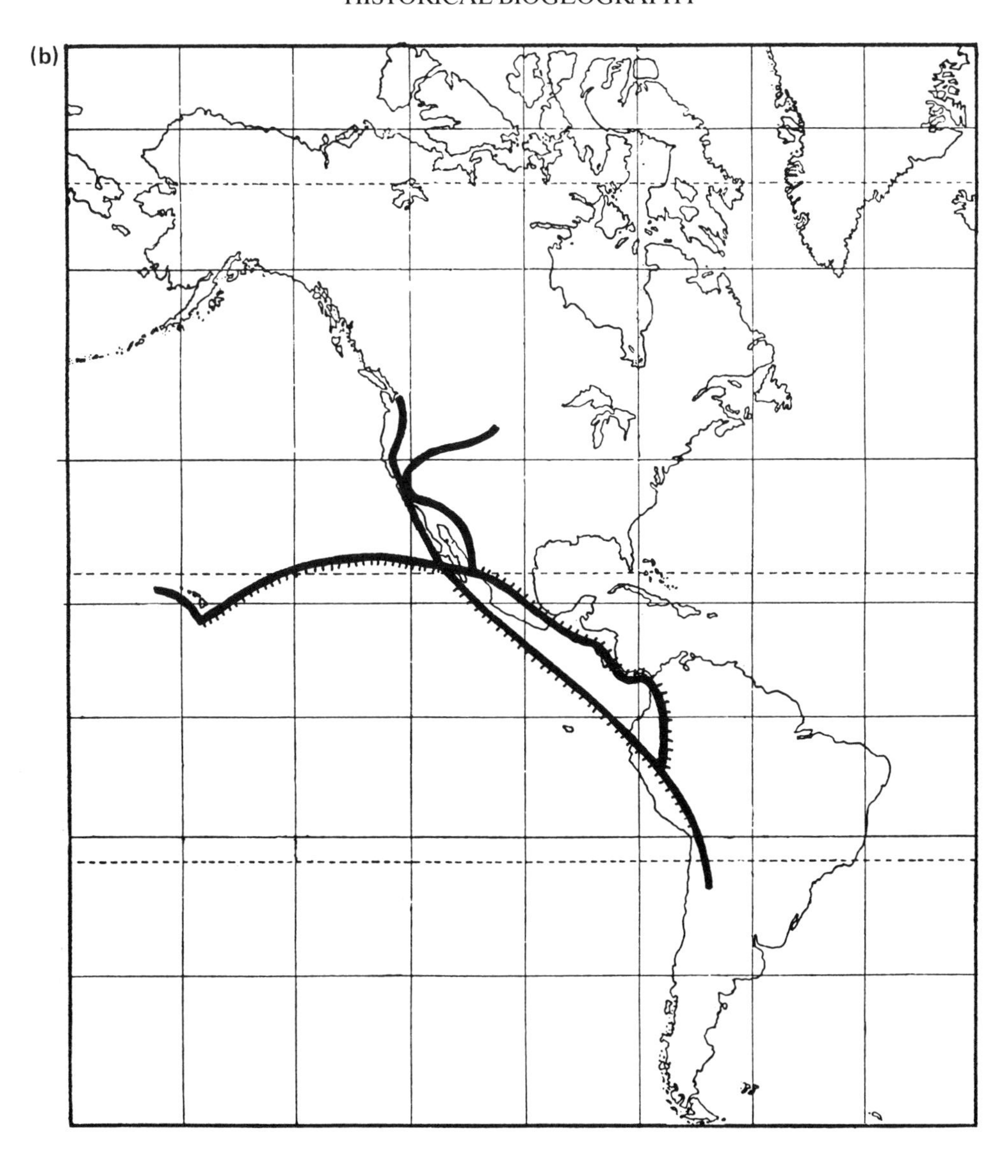

(a) tectonic change, not dispersal, is the causal explanation for different areas containing different species; and

(b) the main biogeographical regions for terrestrial organisms correspond not directly to composite or hybrid modern continents, unlike de Candolle's habitations and Wallace's realms, but rather to modern ocean basins (Nelson 1978*b*, p. 295).

His interpretation was a break from the Darwin–Wallace tradition, and instead a development from the ideas of Buffon, de Candolle, Hooker, and Wulff. Croizat thought that tectonic change was all important, for it is that process which allows 'form-making' or recognizable taxa in particular areas which give us the present day distinctive forms of animal and plant life. A direct corollary is that the distributional areas are causally interrelated, and the basis of that relationship is subject to investigation because all species and their definable distribution areas must ultimately exhibit a global pattern of interrelationships (Nelson 1978*b*, p. 296).

As distinct from dispersalism, which separates geological history and distribution, Croizat's (1958) view in *Panbiogeography* brought the two together. His view on the role of dispersal is novel. For him 'dispersal' is the causal factor to explain related species occupying different areas. Instead of a species migrating from one area to another and fragmenting the range, disjunctions are seen as a result of tectonic change and they represent the ranges of former taxa.

For us, the best description of Croizat's method is in *Space, time and form, the biological synthesis* (1964), but see also Craw (1982). For example, Croizat (1964, p. 7) considered the breaks in distribution of taxa occurring in North and South America. Repetitious gaps or disconnections occur between Mexico and Peru, Mexico and Bolivia, Mexico and Chile, and so on. The areas of endemism of groups of related taxa are connected together by a line or track—which can be indicated on a map as a graph of geographic distribution. In the example (see Fig. 1.9) the track indicates that a certain taxon occurs in western North America, Hawaii, and Bolivia, the last two areas of endemism being sharply separated from the first. In other words, the track places the coordinates of a group in *space* and according to Croizat (1964, p. 7) opens the way to an enquiry into factors of time and form. As we shall see in Chapter 2, cladistic biogeography has re-ordered the dependencies because considerations of form are an essential prerequisite to an enquiry into time and space. Thus, a track becomes the distribution graph of a monophyletic group.

The distribution of one track may coincide with the distributions (tracks) of other taxa. Coincident distributions involving several unrelated species or monophyletic groups (coincident individual tracks) conform to reality and are components of a general biotic distribution or a generalized track (Fig. 1.9). The distribution of most species and of most monophyletic groups coincides with part or all of that of some other species or group and many therefore occupy one part or all of a generalized track. The most generalized tracks include the largest number of, and the most biologically diverse groups of organisms, both fossil and Recent, and are therefore the most thoroughly confirmed (Croizat *et al.* 1974, p. 265). As Croizat (1964, p. 7) noted, if a given individual track recurs in group after group of different organisms to yield generalized tracks, these coincident distributions become statistically and geographically highly significant and require general explanations.

When disjunct monophyletic taxa occur repeatedly in the same areas of endemism, the pattern is likely to be due to one of a series of historical events that created that pattern of disjunction. In other words, generalized tracks represent the present day patterns of ancestral biotas. By determining what major types of coincident patterns occur on the globe today, the number of individual tracks composing each and the variety of organisms incorporated (Croizat 1964, p. 21), the ancestral biotas may be realized (Croizat *et al.* 1974).

Of the better known examples of generalized tracks, we cite Hooker's example of the southern hemisphere areas of South America, Tasmania, Australia, and New Zealand (Fig. 1.10). Different groups of organisms occur in these four southern areas and exhibit identical, or nearly identical, patterns of disjunction. Thus, the near identical distributions of freshwater fishes, earthworms, molluscs, birds, mammals, insects, mosses, and flowering plants (Croizat *et al.* 1974, Croizat 1952; Good 1964, 1974; Brundin 1966; Patterson 1981*a*; Humphries 1981; Craw 1979, 1982) pose a general problem concerning the original distribution and subsequent history of a pan-austral biota. To what does the evidence add up?

The ancestral biota might once have been geographically widespread and later subdivided in relation to disruptive geological events, or each group might have had a 'centre of origin' elsewhere and come to occupy the present day

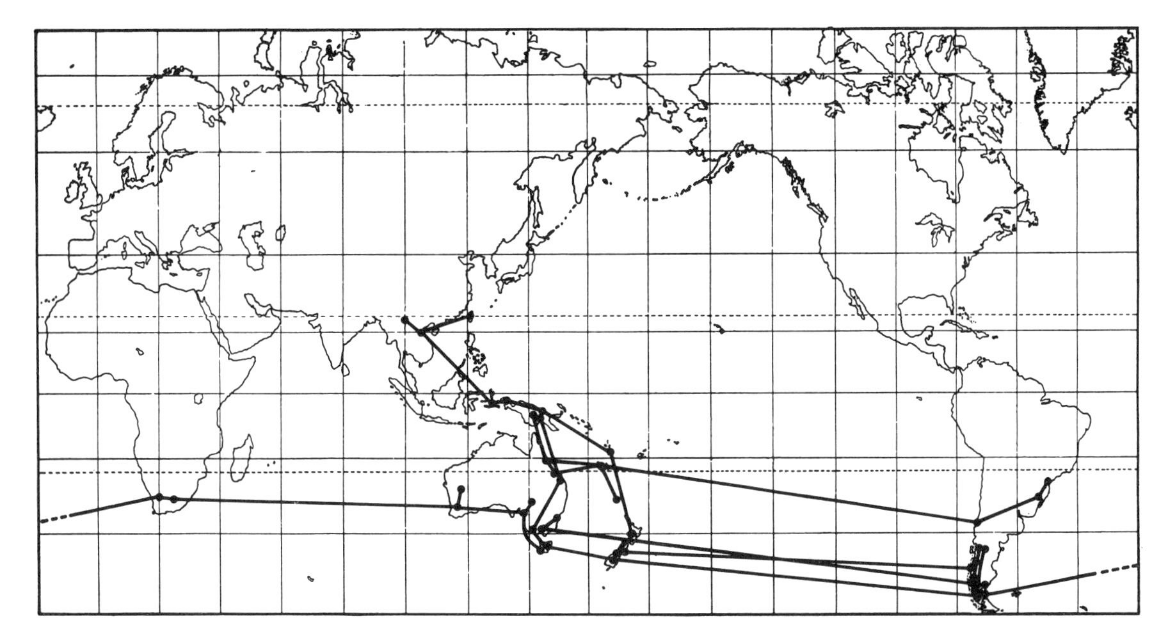

Fig. 1.10. Some generalized tracks for the southern hemisphere and S.E. Asian tropics; Chironomid midges; *Nothofagus*; *Restionaceae*, *Araucaria*, *Libocedrus*.

distributions by migration and colonization. The problem is how one assesses such patterns. Can it really be possible that *Nothofagus* seeds or thousands of tiny midges are carried by long distance dispersal to other lands thousands of miles away? Yet, so many identical tracks occur for many different groups. A generalized track is a constraining reference and it is more likely that the pattern is caused by a single event or a series of historical events to explain the repeated patterns.

Generalized tracks are not restricted to terrestrial examples but may also concern marine biotas that were once formerly continuous and have since subdivided. A spectacular example is the amphi-American marine biota separated by the isthmus of Panama, where several monophyletic groups exhibit an eastern Pacific-Caribbean track.

The advantage of the track and the generalized track are that the co-ordinates in space are determined on the general relationship of taxonomic groups totally independent of geological considerations. The limitation of tracks however, as we shall see in Chapter 2, is that they only connect the areas which are causally related without considering the pattern of relationships of the areas connected by the track (Fig. 1.11, after Croizat *et al*. 1974; Rosen 1976).

1.6 CONCLUSIONS

Historical biogeography, from Linnaeus to Croizat, may be seen as a series of different propositions for the causes of present day distribution patterns. For de Buffon's patterns of allopatric distribution, de Candolle suggested that earth history was all important. For Willdenow and Humboldt, former land-bridges explain disjunctions. The significance of this on a global scale came with the work of Croizat. For Wallace and many twentieth century workers such as Matthew, Cain, and Simpson, dispersal by Darwin's 'occasional means of transport' played a significant role which effectively immunized their efforts from criticism. The problem with all the solutions is that they lacked a testable definition

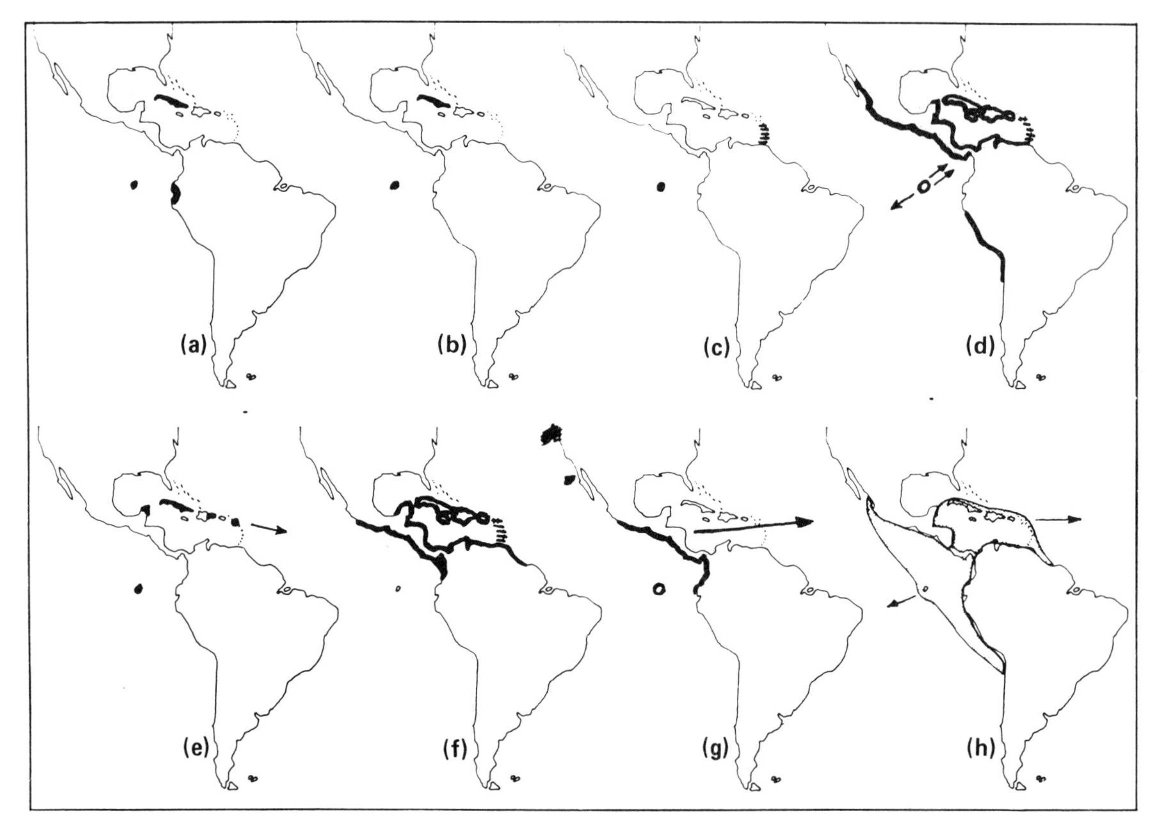

Fig. 1.11. Generalized tracks of monophyletic components of the amphi-American biota (the eastern Pacific–Caribbean generalized track). (a) isopods of the genera *Jimenezia*, *Colombophiloscia*; (b) isopods of the genera *Troglophiloscia* and *Nesophiloscia*; (c) extinct rice rats of *Megalomys*; (d) brachyuran crabs; (e) atyid shrimps of *Typhlataya*; (f) several fish groups; (g) generalized distribution of stomatopod crustaceans; (h) generalized track for all groups in (a)–(f). (After Rosen 1976, Fig. 5, p. 441.)

of relationship. The concept of regions which are compared with each other in terms of degrees of similarity is rather meaningless. Even generalized tracks, although they offer a solution of identifying the problem of areas of endemism, contain only generalized statements of the connections between them. Until Hennig (1950, 1965, 1966) clarified the concept of relationship in systematics the question of relationship between areas of endemism similarly could not be understood.

The lessons from history are clear. For biogeographical theories to be meaningful, falsifiable statements must be derivable from them at two different levels. First, What constitutes an area? Second, What are the interrelationships of those areas? Of the two classes of causal factors, dispersal and vicariant changes, which is the most meaningful for our hypothesis of earth history? The cladistic answers to these questions are outlined in Chapter 2.

2 METHODOLOGICAL DEVELOPMENTS

2.1 INTRODUCTION

The developments in the theory of continental drift and its general acceptance during the 1960s led to the acceptance of the idea that disjunct biotic patterns and corresponding geological patterns were due to the same causes of earth history. The idea of the earth constantly changing means that many, if not all, of the migration solutions to biogeographic problems are wrong. As will be outlined below, a solution came when Brundin (1966) applied Hennig's (1950) definition of relationship to the problems of vicariant distribution of southern hemisphere chironomid midges. The ensuing developments in theory brought Croizat's panbiogeography into the limelight. When fused with cladistics this eventually produced cladistic biogeography. Systematic patterns become understandable and comparable when they are expressed as area cladograms. Area cladograms, branching diagrams of areas, express the interrelationships of areas as determined from systematic information by substituting the taxa for the areas in which they occur (Section 2.3).

Later in the chapter, it will be shown that several different taxonomic groups often show the same pattern of area interrelationships and it is possible to make a general hypothesis about the interrelationships of biotas and areas of endemism. Corroborated hypotheses of this sort can be compared with similarly organized, but independent, information from geology.

2.2 CLADISTICS

Cladistics is a systematic method formulated by the German entomologist, Willi Hennig (1913-1976), alternatively called phylogenetic systematics (from the title of the English translation of Hennig's book, 1966), Hennigian systematics and cladism. A comprehensive review of Hennig's ideas and initial responses to them is given by Dupuis (1978). Hennig's method is significant because in his description of a general method of reconstruction he provided a precise meaning of relationship which triggered off a new vigorous period of systematic research. Hennig's method was originally enunciated for reconstructing phylogenetic trees, but in the last eighteen years has become subtly refined into a method with more general properties and a much wider application than he originally intended.

Similarly, cladistic biogeography has undergone a series of subtle changes which have paralleled changes in cladistic method. Therefore, we give an outline of phylogenetic systematics appealing to common ancestry for definition of monophyletic groups as originally advocated by Hennig (1966), since it is presented in this way by some authors (e.g. Eldredge and Cracraft 1980; Wiley 1981), together with a description of so-called 'transformed' cladistics which emphasizes that pattern analysis be restricted to empirical considerations (e.g. Platnick 1979; Patterson 1980, 1982*a*, *b*; Nelson and Platnick 1981).

Darwin's (1859) theory of evolution by natural selection has two aspects; the pattern of relationships caused by evolution through common descent, and a mechanism of change by natural selection. His theory was intended to explain the hierarchy of animals and plants. Groups of daisies and sunflowers, and foxes and rats, belong to larger groups of flowering plants or mammals and these belong to even higher groups of land plants and tetrapods, all of which evolved from a single common ancestor (Fig. 2.1). Linnaeus and his successors recognized that organisms could be described as species and then arranged into natural groups, such as genera, families, and orders, in which each level of the hierarchy could be categorized by similarities. Linnaeus was dealing with what he considered to be the Creator's

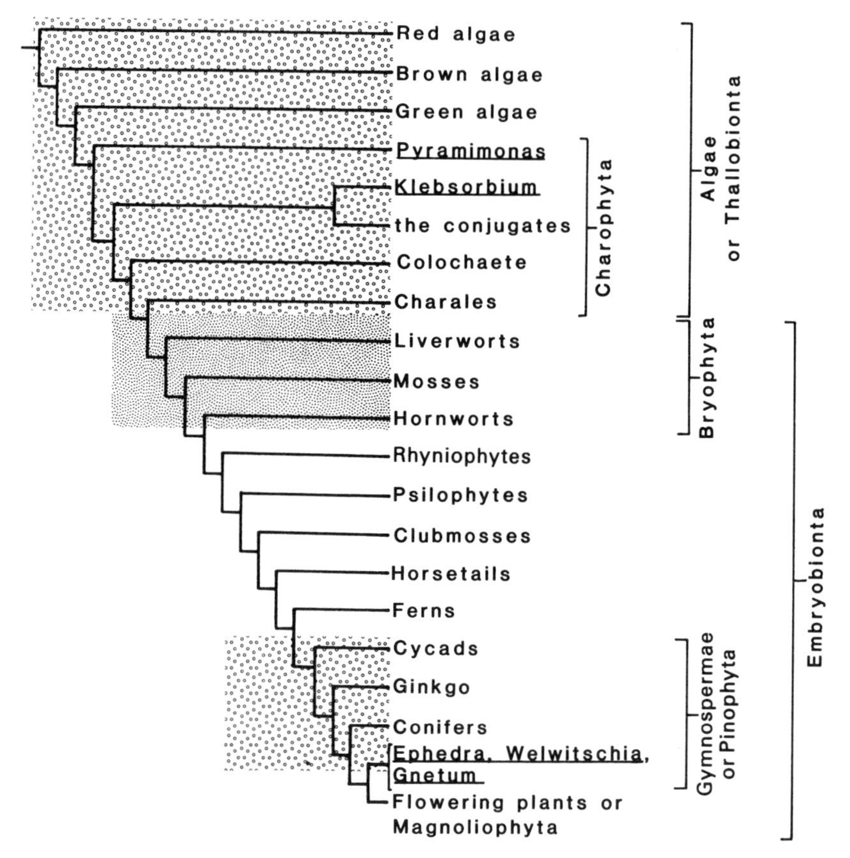

Fig. 2.1. Genealogies and traditional classifications in the land plants and their closest allies. The nodes represent the real groups based on group-defining characters; only one of the taxonomic classes, the Embryobionta, agrees with the genealogy. The shaded areas agree with some of the traditional groupings. (After Hill and Crane 1982; Parenti 1980; Bremer and Wanntorp 1981.)

plan and the grouping characters of genera and families were the essences corresponding to such a plan. Darwin's (1859) contribution was to suggest that the hierarchical relationship between genera and families were 'blood' relationships or kinships caused by descent from a common ancestor. His famous dictum in the *Origin of Species* was 'our classifications will come to be, as far as they can be so made, genealogies; and will then truly give what may be called the plan of creation.'

Darwin's expectations have mostly failed to materialize because classifications are constructed for two purposes: to express phylogenetic relationships and to act as identification keys based on similarities or dissimilarities between different groups. As Patterson (1982*a*) pointed out it has traditionally been understood that these two aims come into conflict because relationships based on common ancestry are almost invariably more complicated than relationships of similarity or differences on which keys are based (see Fig. 2.1). Cladistics is one of the most refined methods that offers a solution to such a discrepancy by tying the phylogeny and the classification into one and the same. The method gives results of a stronger utility for biogeographical problems than non-phylogenetic methods (such as overall similarity methods) which do not try to combine both aspects.

The importance of Hennig's method is his definitions of phylogenetic relationship and his discussion of how relationships are recognized. A phylogenetic relationship can be demonstrated with a rooted, branching diagram; a cladogram (Fig. 2.2). Species B is considered to be more

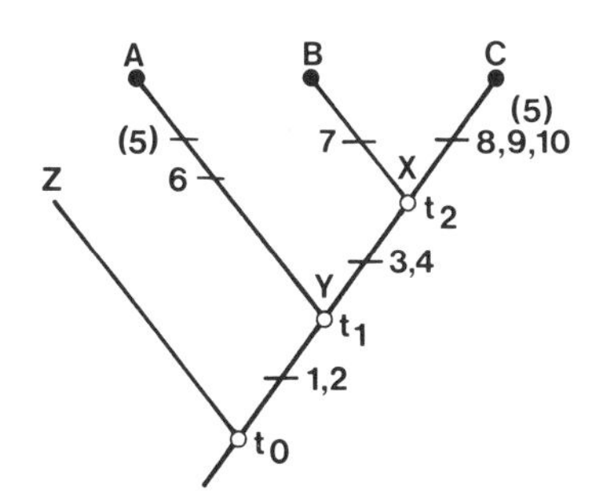

Fig. 2.2. A cladogram showing Hennig's (1966) definition of relationship. A, B, C and Z represent modern taxa; X and Y represent ancestors; numbers represent 1-10 characters (those in parentheses are parallelisms); t_0, t_1, t_2 represent relative time intervals.

closely related to species C than to another species A, since B and C have at least one common ancestor (X) at a later time (t_2) which is not ancestral to species A. Similarly, species A, B and C form a monophyletic group because they have a common ancestor (Y) at t_1 not shared by any other taxon (e.g. Z). Such groups are called monophyletic groups and the task of phylogenetic systematics is to find them. Hennig considered cladograms as phylogenetic trees and monophyletic groups could only be recognized from shared derived characters (which he called synapomorphies) inherited from the most recent common ancestor. Shared primitive characters (symplesiomorphies) inherited from a more remote common ancestor are irrelevant in the search for genealogies. For example, in Fig. 2.2 the relationship of B and C may be obscured by features inherited from Y (at t_1). The bony spinal column and the ability to lay eggs of the duck-billed platypus are uninformative with respect to its relationship with other mammals because both characters (symplesiomorphies) occur in a much larger group, including lizards and birds. On the other hand, hair and mammary glands, for example, are shared derived characters (synapomorphies) unique to monotremes and other marsupial and placental mammals. Characters unique to a group or species Hennig called autapomorphies. Thus he divided the concept of resemblance into three sorts: autapomorphy, synapomorphy, and symplesiomorphy, which describe the status of any character in relation to a particular problem; hair and mammary glands are autapomorphies when one is trying to discover the relationship of mammals to lizards and birds, symplesiomorphies when trying to discover the interrelationships of rats, sheep and pigs, and synapomorphies when trying to discover the relationship of a duck-billed platypus to a rat and a bird.

From concepts of character distribution, Hennig derived definitions for three types of groups (see Fig. 2.2).

1. Groups based on resemblance due to convergent or independently derived characters (e.g. character 5), not inferred to have occurred in the common ancestor (AC—Fig. 2.2); a group containing bats and birds based on a character stated as 'possession of wings' would be one example.

2. Monophyletic groups based on shared resemblance derived from a common ancestor. In Fig. 2.2 the group ABC based on characters 1 and 2 in ancestor Y and the group BC based on characters 3 and 4 in ancestor X are both monophyletic groups. Vertebrates based on backbones, mammals based on hair and viviparous mammals based on live birth are all examples of monophyletic groups.

3. Para- and polyphyletic groups, i.e. those that do not contain all the descendants of a common ancestor. A paraphyletic group is one remaining after one or more parts of a monophyletic group are separated because the members share many derived characters. Thus, in Fig. 2.2 taxon C would be put in a group of its own because it has three times as many unique characters (8, 9, and 10) as either A or B with one unique character

each (6 and 7 respectively); the group AB is paraphyletic. It is a group without a single defining character recognized only by the absence of characters 8, 9, and 10. Classic examples include the Gymnospermae or the Algae amongst plants (Fig. 2.1) and Reptilia or Invertebrata amongst animals. There is no uniquely derived character that defines any of these groups.

Hennig's work was to find phylogenetic relationships of various dipteran taxa, and his method was set in an evolutionary framework. His definitions of relationship, monophyly, synapomorphy, and symplesiomorphy are based on a notion of relative recency of common ancestry and transformation of characters during evolution. Cladograms were (to Hennig) absolute phylogenetic trees with the nodes, or branching points denoting divisions of ancestral species and an implicit time axis from the most inclusive dichotomy to the terminal taxa (Fig. 2.2). However, it is possible to see cladograms in a more generalized framework—with a relative timescale and the nodes representing shared derived characters (synapomorphies) without the notion of a particular pattern of ancestry; instead the cladogram is the pattern of character distributions in the natural hierarchy. In a phylogenetic tree the nodes, instead of representing the defining characters, are ancestors, the branch points speciation events, and the lines actual lineages of descent with modification (Platnick 1979; Patterson 1982*a*, *b*). By contrast a cladogram is a general expression of empirical evidence; of organisms and their characters consistent with a variety of phylogenetic trees. Figure 2.3(a), the cladogram, if viewed as a tree, says that species A, B and C are all alive today, that the two hypothetical ancestors X and Y are now extinct and that speciation was strictly dichotomous. The five other trees (Fig. 2.3(b)-(f)) are all equally consistent with the four grouping characters (1-4) but give other interpretations for the precise course of history. The cladogram (Fig. 2.3(g)) can thus be interpreted as a summary of a general pattern without any particular evolutionary history specified by invoking ancestors. It is therefore, only a generalized evolutionary tree and indeed the only one accessible on the empirical evidence at hand.

The principal task for cladists centres on the discovery of monophyletic groups. The cladistic method demands that for groups to be considered monophyletic, they must be characterized by synapomorphies (homologies). Many evolutionary taxonomists consider a major disadvantage of cladistic methods to be that evolutionary divergence is not weighted, especially in classifications. By way of an example consider the relationships of the flowering plants with other land plant groups, such as *Welwitschia*, *Ephedra* and *Gnetum*, the conifers and cycads (see Parenti 1980; Hill and Crane 1982, Fig. 2.1). The angiosperms are usually separated off because they have several autaopomorphic characters. The remaining taxa are classified as the gymnosperms. The problem is that the Gymnospermae is not a group by definition but by default. A diagram of the relationships based on an evolutionary interpretation disagrees with a cladogram based on character analysis, and in fact has not a single character for its definition. If there is any criterion by which to judge a classification it is by agreement with the greatest number of characters at hand. The best independent test of a cladogram on a given set of characters is to see if an independent set of characters gives a similar cladogram; for example, DNA data and cytological data, as compared with morphological data.

Homologous characters are the only available clues to relationship and as we have seen already these allow us only to make general statements about common ancestry, not specific statements about direct ancestry. Shared characters do not allow us to distinguish between two different stories of ancestry and so other criteria have to be introduced. Fossils are most frequently used for interpretations of ancestry, especially when certain fossils appear earlier in time than others. However, even with the best stratigraphic sequences, fossils have to be interpreted in the same way as modern taxa. Fossils are invariably incomplete and only interpretable by comparison with modern taxa. Palaeontologists faced with the

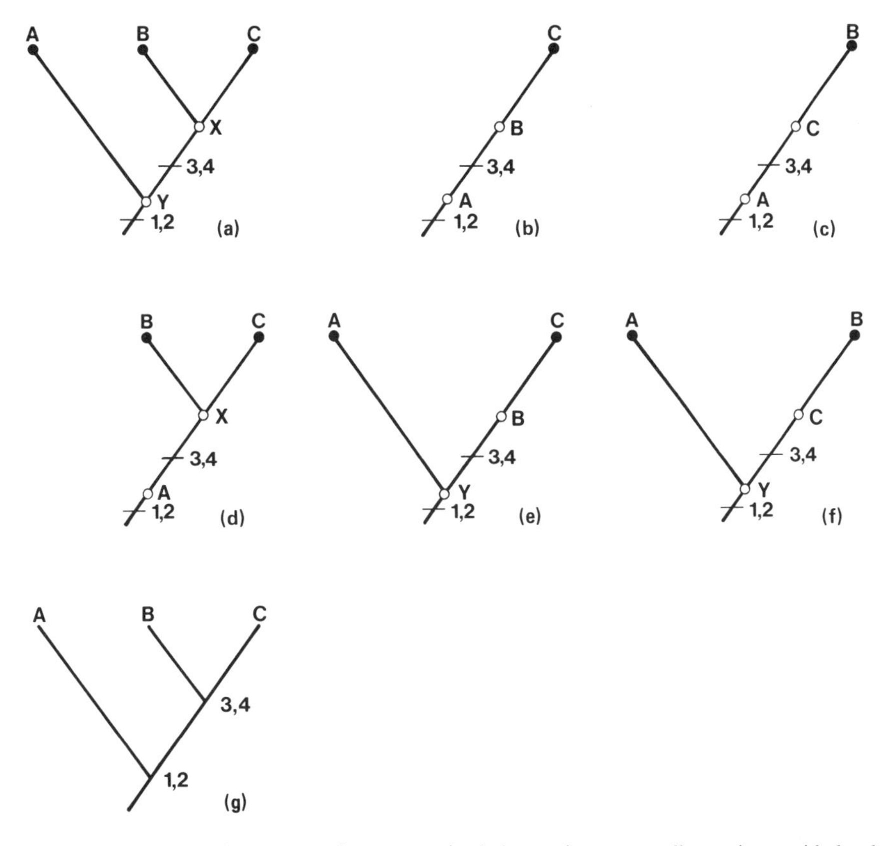

Fig. 2.3. Cladograms and phylogenetic trees; (a)-(f) represent six phylogenetic trees equally consistent with the cladogram in (g) defined on the basis of four characters (1-4).

problem of naming a particular ancestor for a particular group will name instead ancestral groups—the Pro-gymnospermopsida being ancestral to Gymnospermae—as an approximation to the idea that, if found, the actual ancestor would belong to that group. Extinct ancestral groups are always paraphyletic, just like the Gymnospermae, Reptilia and Magnoliatae (dicots), and are just artefacts in the same way. Because paraphyletic groups cannot be identified by real characters—or only by primitive characters (e.g. naked seeds for the Gymnospermae) there are no limits to their formation. Seen in this light statements like the 'Gymnosperms evolved from the Pro-gymnospermopsida', (see Beck 1976) seem absurd since it is fiction to say that one non-group evolved from another non-group. Thus, cladistics allows us to discard another conventional aspect of dispersal biogeography—that it is necessary to have fossils to reconstruct the distributional history of a group.

2.3 CLADISTICS AND BIOGEOGRAPHY

The earliest applications of cladistics to biogeography were attempts by Hennig (1966) and others to use cladograms to determine the 'centres of origin' of monophyletic groups. Hennig (1966) for example, reasoned that there is a close relationship between species and the space each one of them occupies. But rather than saying that species and spaces evolve together, he assumed instead that dispersal patterns are unique for each taxonomic group and each has an independent history.

The best applications of Hennig's method are found in Brundin (1966, 1972*a*, *b*, 1981) for chironomid midges and also Ross (1974) for caddis flies (Section 2.3.1).

Although Hennig, Brundin, and Ross brought much greater precision to biogeography by superimposing areas onto phylogenies and inferring the lowest number of dispersals for each group, the method relied on *ad hoc* assumptions that groups have 'centres of origin' and species migrate. The breakthrough in the application of cladistic reasoning to biogeography came, in our belief, with the efforts of biogeographers such as Nelson and Rosen in their interpretation of Croizat. Instead of a 'vacuum' theory of biogeography, whereby certain areas were originally devoid of taxa later to be colonized from other source areas, there was proposed an equally plausible alternative. Disjunct distributions could come about by vicariance events because their ancestors originally occurred in the areas where they occur today, and the taxa evolved in place (Croizat *et al*. 1974). In other words, dispersal models explain disjunctions by dispersal across pre-existing barriers, vicariance models explain them by the appearance of barriers fragmenting ancestral species ranges. So what became particularly clear was the important idea that distributional data are insufficient to resolve decisively either dispersal or vicariance as the cause of a disjunct distribution pattern. Therefore, when faced with a particular distribution as Platnick and Nelson (1978) argued, one should not worry about its cause but whether or not it conforms to a general pattern of relationships shown by other groups of taxa endemic to the areas occupied. Thus, as in cladistics where three-taxon statements are the most basic units for expressing relationships—a cladogram indicating relative recency of common ancestry—in biogeography three-area statements are the most basic units for expressing relationships—a cladogram indicating relative recency of common ancestral biotas. Area cladograms are produced by substituting for taxa the areas in which they occur. The generality of the area cladograms can be examined by comparison with other unrelated taxonomic groups endemic to the relevant areas and corroboration of a particular pattern is equivalent to a general statement for the relative recent ancestry of the biotas under examination.

Initial applications of the method (e.g. Rosen 1976) encountered problems of incongruence and unresolved statements in the general area cladograms (Section 2.3.2). Theoretically, it should be possible to connect every area of endemism into one larger general statement of interrelationships. However, our perception of the world is less than perfect for a variety of reasons—extinction, dispersal of widespread taxa, and restricted distributions of taxonomic groups.

2.3.1 The progression rule

Central to Hennig's (1966) method to find a 'centre of origin' for a group of taxa from a particular cladogram, was the idea that phylogenetically primitive members of a monophyletic group will by definition be found near that centre. In other words, within a continuous range of different species of a monophyletic group it was considered possible in certain circumstances that a transformation series of characters would run parallel with progression in space, such that the youngest members would be on the geographical periphery of a group. A good example is given by Ross (1950) with the *Wormaldia kisoensis* complex of caddis flies. The geographical distribution of the nine species, eight species in the Asian Pacific between Sarawak in the south and Japan in the north and one species in the Smoky mountains of eastern North America, is shown in Fig. 2.4. The cladogram of phylogenetic relationships is shown in Fig. 2.5. Ross (1974) assumed that the base of the stem denotes the ancestor of the whole group. That the most derived species pair occur in Japan and eastern U.S.A. led Ross to consider the simplest dispersal hypothesis of an origin for the

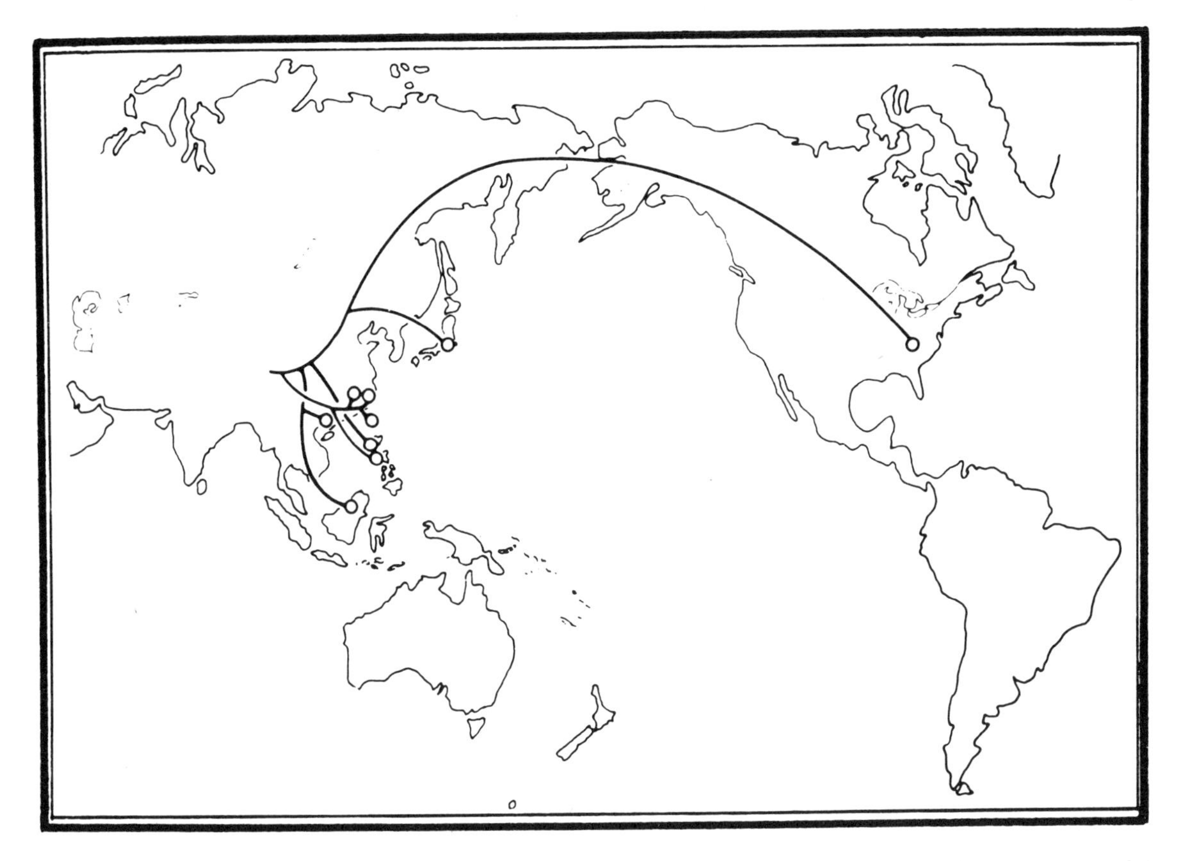

Fig. 2.4. Distribution and phylogeny of *Wormaldia*. The circle in Japan is *W. kisoensis*, that in eastern N. America *W. mohri* (from Ross 1974, p. 217.)

group in Asia and a single dispersal event for one species across the Bering straits to North America.

Brundin's classic studies of chironomid midges (Brundin 1966, 1972*a*, *b*, 1981) showed that the southern temperate areas of South America, Southern Africa, Tasmania, south-east Australia and New Zealand are inhabited by 600 to 700 species. Transantarctic relationships are a recurring phenomenon throughout the group, so by way of an example comments will be restricted to the cold running water midges of the subfamily Diamesinae, which display a double extra-tropical distribution: two major groups present in both northern and southern hemisphere temperate areas but absent from the tropics. The largest, and most widespread is the relatively generalized tribe Heptagymi, represented by eleven

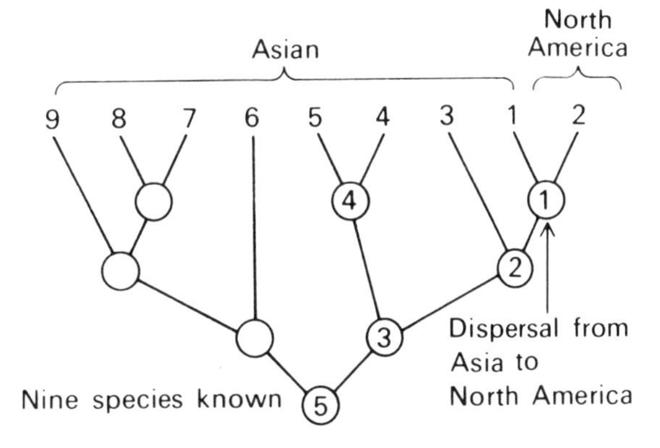

Fig. 2.5. Cladogram for the nine species of *Wormaldia* (see Fig. 2.4). (From Ross 1974, p. 216.)

species in Andean South America, two species in south-eastern Australia and five species in New Zealand. Its sister group is the relatively more apomorphic and monotypic tribe Lobodiamesini of New Zealand. The cladogram in Fig. 2.6 shows that there are a total of 25 terminal taxa in the southern hemisphere areas of South America, New Zealand, Australia and South Africa, and three groups in Laurasia. The monotypic genus *Heptagyia* occurs in South America and *Paraheptgyia* has five South American species. According to Brundin the south-eastern Australian subgroup of two species is a younger evolved offshoot of the older South American group including *Heptagyia*. Brundin considers the Australian taxa to have dispersed from Patagonia or east Antarctica at stem 6 by the end of the Palaeocene because they have derived characters and because the stem species (indicated by 1, 2, and 4) never occurred in Australia. The other stem (2a) includes *Reissia* with three species in South America, *Limaya* with two species in South America, and *Maoridiamesa* with five species in New Zealand. Brundin (1981) considers the fact that *Maoridiamesa* is on a different stem from the Australian *Paraheptagyia* group agrees well with plate tectonic theory for an early separation of New Zealand from western Antarctica in the Upper Cretaceous. The fact that *Maoridiamesa* is a comparatively younger, derived offshoot of an older group in South America is, according to Brundin, evidence of long distance dispersal from South America via west Antarctica to New Zealand of stem species 4a rather than a vicariance event. In other words, Brundin thinks that the *Maoridiamesa* group is younger than the areas in which it occurs.

Interpreting cladograms as phylogenetic trees rather than as hierarchical groups with relatively more inclusive groups of set membership, often requires making *ad hoc* assumptions not fully justified by the information on which they are based. Furthermore, interpreting individual cladograms as having individual histories leads to certain conceptual difficulties. A crucial one is the repetition of distribution patterns. If we have, as is this case, many unrelated taxonomic groups repeating a pattern of distribution between major continents such as South America and Australia and New Zealand, it is improbable that there were

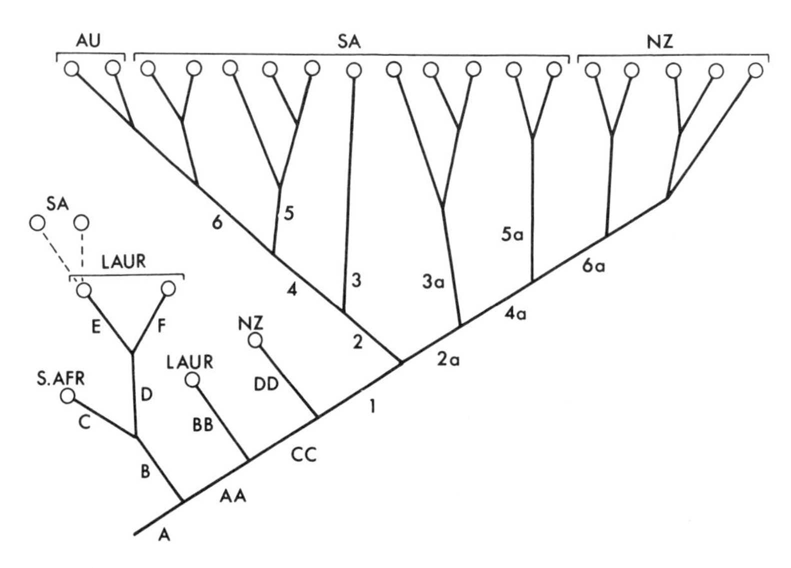

Fig. 2.6. Partial reconstruction and area cladogram of the subfamily Diamesinae (Diptera; Chironomidae) 1, Heptagyini; 3, *Heptagyia*; 4, *Paraheptagyia*; 3a, *Reissia*; 5a, *Limaya*; 6a, *Maoridiamesa*. A, Diamesinae; C, Harrisonini; E, Diamesini; F, Protanypodini; BB, Boreoheptagyini; DD, Lobodiamesini. (From Brundin 1981, Fig. 3.7, p. 119.)

many dispersal events, with each group separately making its way from one continent to the other. The most logical and simplest conclusion would be to suggest that at one time, the continents were in contact and that the present day pattern was due to the break up of a formerly continuous biota.

2.3.2 Vicariance biogeography—Rosen's method (1976)

Rosen set out to apply Croizat's vicariance method to Caribbean biogeography. The application has special significance because it was the first concise exposition of panbiogeography with cladistics added, so that the groups considered are monophyletic rather than both monophyletic and paraphyletic, as in many of Croizat's examples. The method consisted of marking the distribution of disjunct components of monophyletic groups on a map, and linking the areas of each group by a line or track. Where tracks linking sister taxa (fossil or Recent) coincided repeatedly, the lines delimit a generalized track, assumed to link two or more biotas that are vicariant products or fragments of a single ancestral biota. The sequence of events was compared to the data from historical geology to account for the fragmented pattern.

Rosen recognized four generalized tracks for the Caribbean region: A North American–Caribbean track and a South American–Caribbean track, both mainly terrestrial; and an eastern Pacific–Caribbean track and an eastern Atlantic–Caribbean track, both mainly marine (Fig. 2.7). His sources of distributional data include plants, amphibians, reptiles, birds, mammals, and fish.

The main problem was to determine what these tracks mean in terms of distributional history. Rosen suggested that the four generalized tracks may be interpreted either as a result of four separate dispersal routes into the Caribbean areas or as vicariance events that subdivided ancestral, marine, and terrestrial biotas into smaller units. The first interpretation demands that dispersal was remarkably co-ordinated for such vastly different groups as birds, plants, and amphibians. The vicariance interpretation required geological events that isolated the eastern and western Atlantic, and the eastern Pacific from the Caribbean followed by an intermingling of North and South American biotas in the Mesoamerican region.

Rosen (1976) regarded dispersal hypotheses as untestable because they appeal to individual explanations: 'Thus, dispersal theories, if they attempt to deal at all with distributions in a rigorous way, are rather complicated and incorporate a major unexplained ingredient—namely the co-ordinated movements via active migration and chance dispersal of countless organisms of vastly different biological properties.' (Rosen 1974, p. 445). Vicariance hypotheses on the other hand can be tested in two different ways: either to add further individual tracks, or to compare the biogeographical hypothesis with a geological one. In the first test, new tracks will either be congruent or incongruent with the generalized track. The test is difficult to put into practice because incongruent tracks, assuming the taxonomy to be correct, can also be interpreted in one of two ways: as belonging to different generalized tracks or as dispersal events. Rosen pursued the second type of test in detail and found very close agreement between the four-track model and Malfait and Dinkelman's (1972) account of Caribbean history.

The sequence of events for Rosen's general hypothesis which combined biogeography and geology was as follows (after Patterson 1981*a*):

1. Development of a later Jurassic proto-Antillean archipelago linking North and South America, colonized by dispersal from both continents and allowing limited dispersal of each continental biota to the other land mass.

2. Displacement of the proto-Antilles to the east, as Pacific sea floor intruded between North and South America (the archipelago would have carried a mixed and now isolated North and South American biota, and the eastern Pacific marine biota would not disperse into the Caribbean).

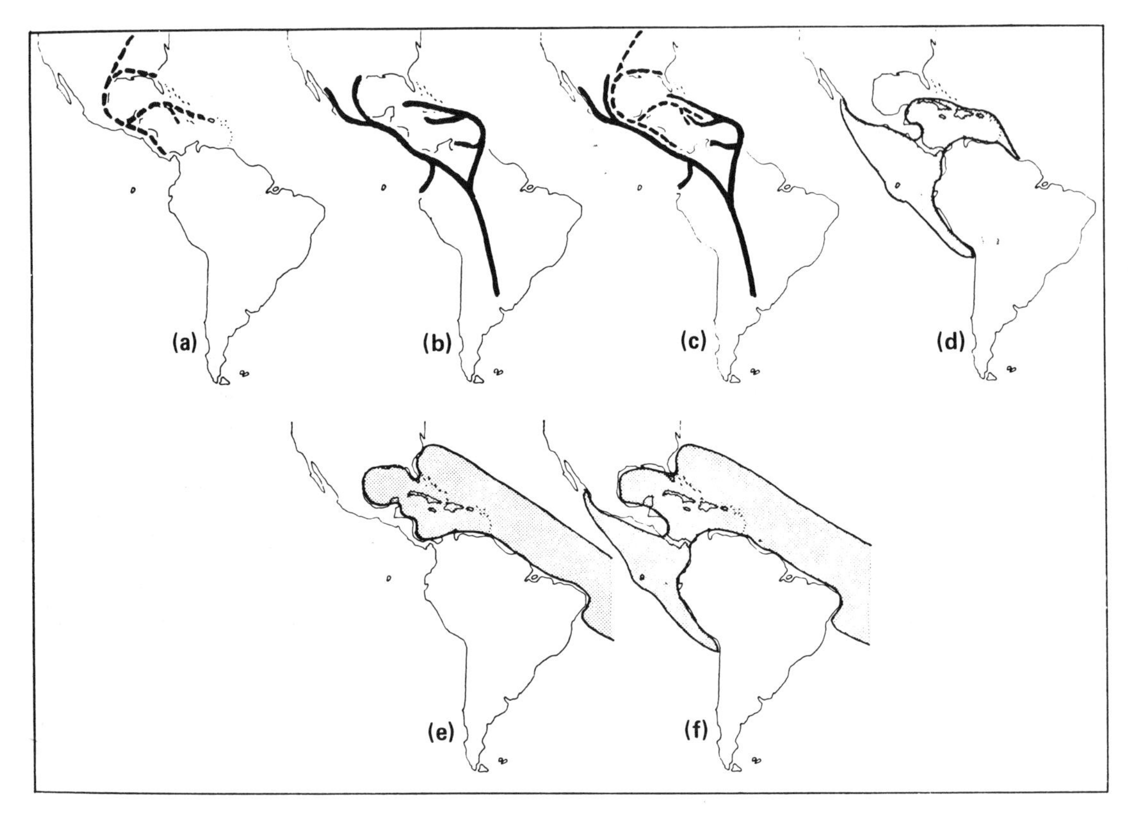

Fig. 2.7. Summary of transcontinental and transoceanic generalized tracks. (a) North American Caribbean track; (b) South American Caribbean track; (c) overlapping of North and South American Caribbean tracks enclosing the Caribbean sea; (d) Eastern Pacific Caribbean track, the trans-oceanic track hypothesized to be youngest depicted here; (e) Eastern Atlantic western Atlantic track, a track of intermediate age; (f) Eastern Pacific eastern Atlantic track, hypothesized to be the oldest transoceanic track depicted here. (From Rosen 1976, Fig. 6, p. 444.)

3. Development of an epicontinental seaway that isolated eastern and western North America.

4. Development of a new lower Central American archipelago that partially isolated the Caribbean from the eastern Pacific and was again populated by dispersal from north and south, allowing a second phase of dispersal between the North and South American biotas.

One flaw with this conclusion was the combination of vicariance and dispersal as influenced by geological theory. The striking concordance between biology and geology gives a compelling story. As a method it is not new since classic dispersalist hypotheses (e.g. those of Wallace 1876; Matthew 1918; Darlington 1965) were also found to be concordant with geological hypotheses. As Patterson (1981*a*, p. 455) notes, recent accounts of Caribbean and eastern Pacific geology, especially tectonic sequences, are in disagreement with Malfait and Dinkelman's theory (e.g. Sykes *et al*. 1982). But, even if the geological hypothesis fades away, it does not falsify Rosen's biological data. The generalized tracks of monophyletic groups still stand and still require an explanation.

There is another problem, however. Can a track be refuted in the same way as a cladogram? Since a track merely connects a set of areas defined by a monophyletic group, is the track a meaningful way of expressing relationship in the same sense as cladistics? It appears that generalized tracks are only a statistical measure of similarity between disjunct biotas (Ball 1976). Therefore, to make a comparison with systematics, tracks mean the same as measures of overall similarity as in phenetics (Patterson 1981*a*). But if monophyletic groups are used in the formulation of generalized tracks the latter cannot be the same as similarity measures in phenetics because 'absence' data are disregarded. However, the question remains—are tracks measures of commonality or a measure of hierarchical relationship? Craw (1982) notes that Croizat (1952, 1958) invariably joins more than two areas together and says that they are more than phenetic. 'In point of fact these diagrams (tracks) contain a natural concept of biogeographic regions . . .' (Craw 1982, p. 306.) Since a track is defined on the connections of related taxa between disjunct areas they do to some extent represent a similar generalization in geography as cladograms do in systematics. Tracks as used by Croizat (1952, 1958, 1964), although usually monophyletic, are not necessarily so. Since they are not hierarchical, but minimal spanning graphs connecting nearest neighbours, they show a more general resolution of geographical patterns than cladograms. However, as we show later, they not only indicate the relationships of areas, they especially show the composite nature of existing areas, as do incongruent cladograms.

To improve the level of resolution it seems that what is really required is the same type of information in geology and areas as used in cladistics. In other words, to make satisfactory comparisons between organisms and areas it is necessary to have cladograms of areas that can be compared with taxonomic cladograms—a technique less generalized than track analysis.

2.3.3 Cladistic biogeography—the method of Platnick and Nelson (1978)

The solution to a definition of area relationship came from the method of cladistic biogeography proposed by Platnick and Nelson (1978), which was applied to real examples by Rosen (1978, 1979). The method combined Hennigian cladistics with Croizat's panbiogeographic method, by making cladograms of individual taxonomic groups occupying the same areas. General area cladograms could be derived by adding the individual cladograms together to give statements about biotas.

Platnick and Nelson (1978) began their paper by asking the question 'Why are taxa distributed where they are today?' They gave two possible answers: either they evolved there or they evolved elsewhere and dispersed into new areas. The difference between vicariance and dispersal lies in the relationship between the age of a taxon and the age of the barriers limiting the area. Vicariance predicts that taxa in two (or more) areas and the barriers between them are the same age; whereas dispersal always predicts that the barrier predates the taxa. By way of an example, consider the American ivory nut palms (*Phytelephas*) which occur on both sides of the Andes of north-west South America in lowland rain forest. A probable explanation for their present day distribution is that the Andean mountain chain grew up during the Tertiary and subsequent evolution led to species differences on the east and west sides. In other words, the ancestor of the present day species was older than the barrier but the *separation* into two areas is the same age and due to the formation of the barrier. Dispersal interpretations are thus not easy to test because they are necessarily considered separately for each group of organisms. In fact, many biogeographers believe that each group has its own dispersal history and are not therefore comparable (e.g. Koopman 1981). Vicariance hypotheses can be tested by other taxonomic groups, fossil or Recent, that occur in the same areas under examination, which should have been affected by the same barriers.

The method of analysis involves first of all finding monophyletic groups with taxa occurring in at least three or more similar areas. Cladograms are produced for each group of organisms. The names of the taxa at the terminal tips of the cladograms are replaced by the names of the areas in which each taxon occurs. These are area cladograms, and the tips of the branches are areas. The sum of the areas on one cladogram is equivalent to a track. The sums of similar areas on several cladograms are equivalent to generalized tracks. To obtain a cladogram of biotas it is necessary to add the individual cladograms together. By way of an example, consider three areas of endemism, southern South America, Australia, and New Zealand, in which occur a group of freshwater fishes A, B, and C, and a group of flowering trees X, Y, and Z (Fig. 2.8(a)). The characters of each group are analyzed and cladograms produced (Fig. 2.8(b), (c)). The area of each taxon is then substituted onto the cladogram to give identical area cladograms (Fig. 2.8(d)). The area cladograms are then compared. In this case, we see that they are congruent. The hypothesis for the three areas in this hypothetical example is that Australia and South America share a more recent history than either do with New Zealand in terms of these two groups.

The success with which we find a congruent vicariant pattern in nature depends upon the frequency with which common factors affect the evolution and distribution of two or more groups of organisms. To find congruent patterns Rosen (1978) considered it necessary to delete unique, unresolved, or incongruent components from cladograms. By way of an example, consider the following five areas of endemism—Australia, New Zealand, southern South America, New Guinea, and Africa—and the two monophyletic groups of fishes (A-D) and flowering trees (W-Z; Fig. 2.9(a)-(c)). The area cladograms (Fig. 2.9(d), (e)) are not completely congruent since the fishes have a widespread species in Africa and New Zealand, suggesting that Africa and New Zealand constitute a single area. The flowering trees, by contrast, have one species in each of Africa, New Zealand, Australia, and Papua New Guinea, but not in South America. By deleting either the unresolved

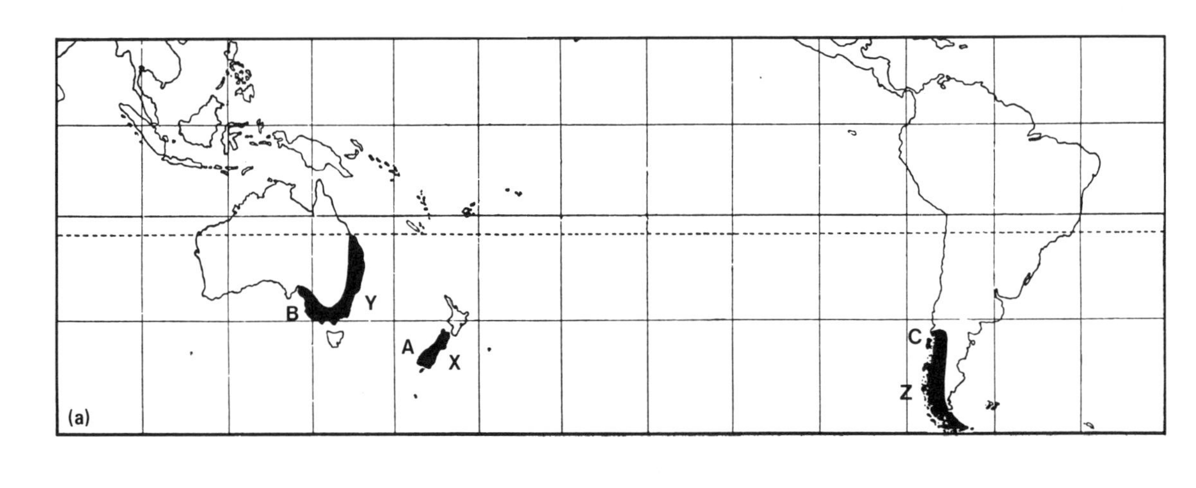

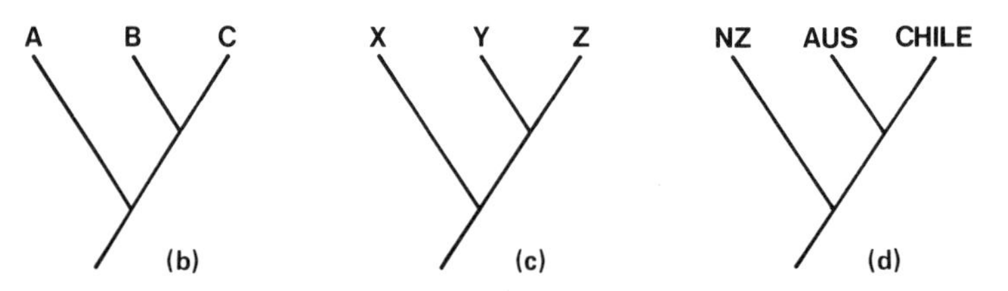

Fig. 2.8. (a) Hypothetical distribution of two groups—fishes (A, B, C) and flowering trees (X, Y, Z). (b, c) Cladograms for each group. (d) Area cladogram common to both groups.

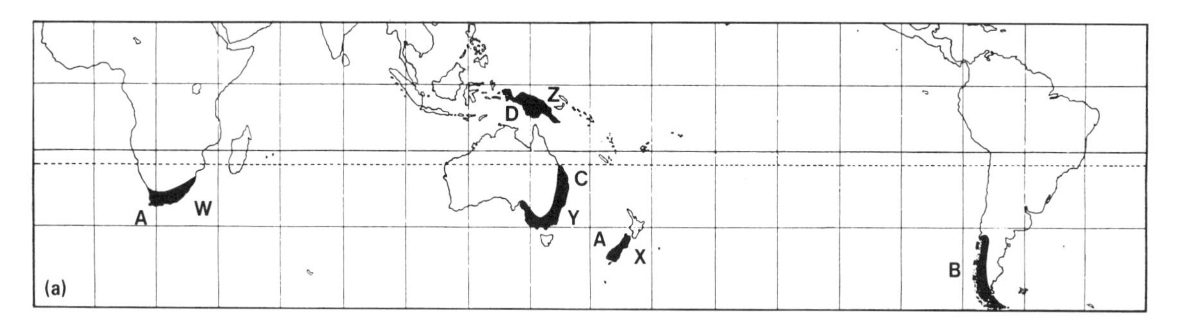

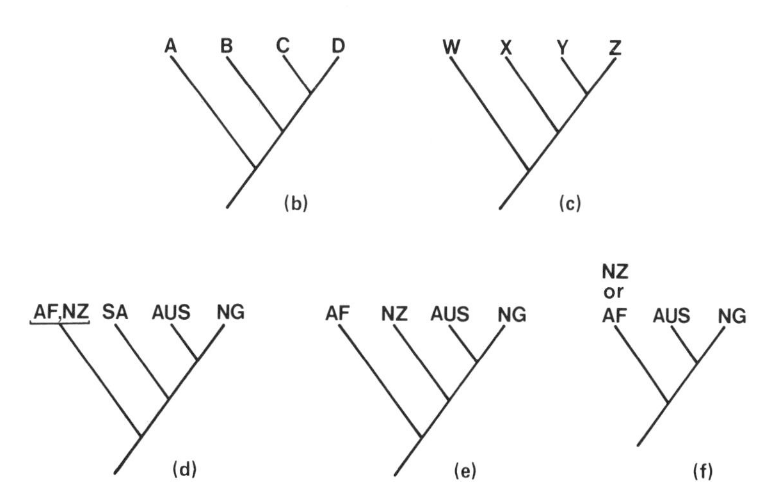

Fig. 2.9. (a) Hypothetical distribution of two groups—fishes (A–D) and flowering trees (W–Z) in the southern hemisphere. (b, c) Cladograms for each group. (d, e) Area cladograms for each group. (f) reduced area cladograms.

African or New Zealand component from both area cladograms and the unique South American component from the fish cladogram we produce two possible reduced area cladograms showing a common three-area pattern with New Zealand or South Africa sister to Australia and New Guinea for both groups of organisms (Fig. 2.9(f)).

2.3.3.1 Poeciliid fish in Middle America—Rosen's example (1978)

The application of the cladistic biogeographic method is still in its infancy because cladistic analyses of plants and animals have only been carried out in a few groups. At least two and preferably more phylogenetic analyses on unrelated taxa must be available for the same set of areas. Rosen's analysis of poeciliid fish in Middle America was the first practical example of applying the Platnick and Nelson (1978) method to real organisms.

Poeciliids are fishes known as killifishes, distinguished by their unique system of internal fertilization; they give birth to fish in an advanced state of growth (see Rosen 1978; Wiley 1981). Rosen (1978, 1979) examined two Middle American poeciliid genera, the platy-fishes of *Heterandria* (Fig. 2.10) and the swordtails of *Xiphophorus* (Fig. 2.11). Both genera have close relatives elsewhere but each has a monophyletic sub-group inhabiting the same eleven general areas in southern tropical Mexico, south to eastern Honduras in *Xiphophorus* and further

Fig. 2.10. *Heterandria cataractae* Rosen. (a) Male; (b) female. (After Rosen 1979, Fig. 12.)

Fig. 2.11. *Xiphophorus cortezi* Rosen. Left—male; right—female. (After Rosen 1979, Fig. 39).

south to eastern Nicaragua in *Heterandria* (Fig. 2.12). Cladograms for the two genera are shown in Fig. 2.13.

Rosen (1978) disregarded area 11 since it was considered to be an area of intergradation between two species in each group. The cladograms for both genera are converted into area cladograms by substituting the area inhabited by each species (Fig. 2.14). Next, simplified cladograms are produced allowing only one term for each area (Fig. 2.15(a), (b)). A reduced area cladogram common to both groups is then produced by deleting the unique information from each cladogram (Fig. 2.16). According to Rosen, unique information does not contribute to the shared history. The common pattern was then inferred to reflect the history shared by *Xiphophorus* and *Heterandria*.

Rosen (1978, 1979) used reduced area cladograms to generate a biogeographical hypothesis for *Xiphophorus* and *Heterandria* because the individual cladograms were incongruent for certain areas. Cladistic biogeogrpahy would be uncomplicated if all groups of organisms were each represented by one, and only one, taxon in each of the smallest identifiable areas of endemism, but this is not the case. Unique components in individual cladograms occur for a variety of reasons: failure of a population to divide in response to the formation of a natural barrier, dispersal from one area to another and extinction in one or more areas, and so on. The problem with Rosen's reduced area cladograms is that they delete data. By deleting the non-congruent elements of different patterns one may be 'fixing' the result. Unique patterns may be meaningful and

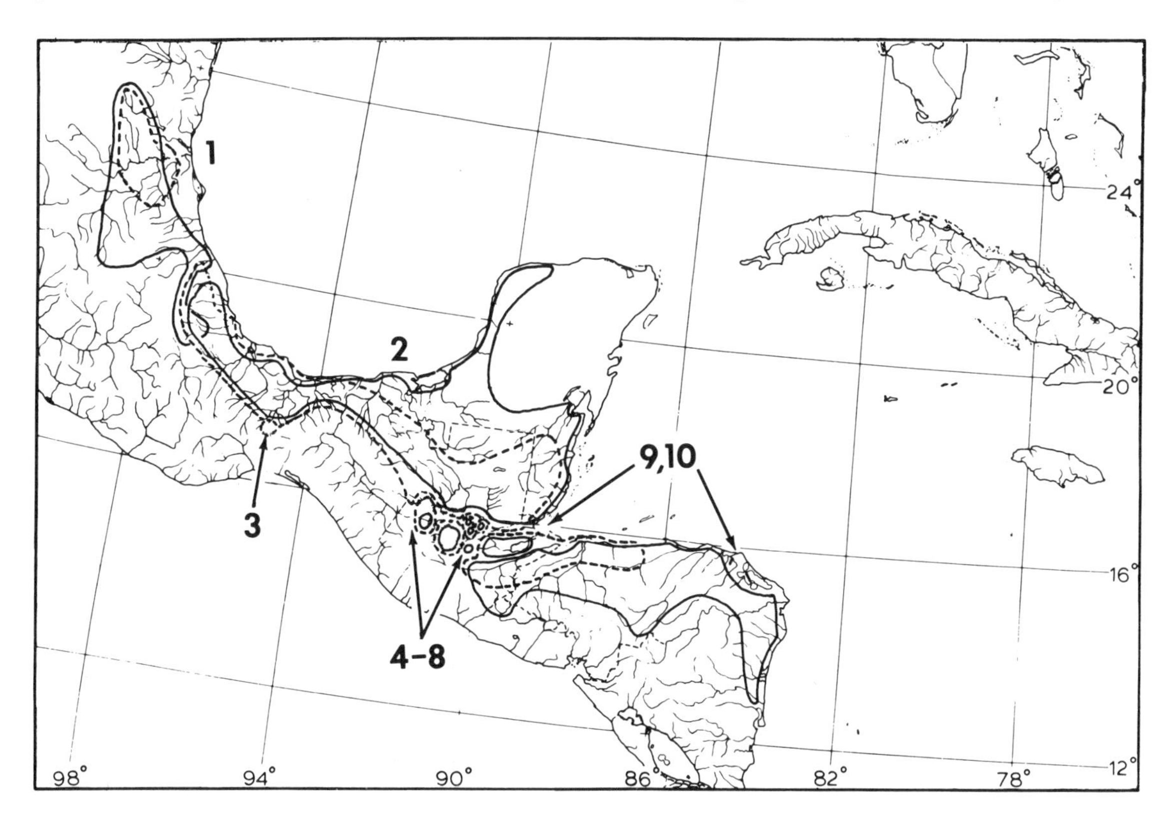

Fig. 2.12. Comparison of the distribution of the species of *Xiphophorus* (dashed) and *Heterandria* (solid) in Middle America. Numbers refer to areas defined by the occurrence of taxa. See text for explanation. (From Rosen 1979, Fig. 45, p. 367.)

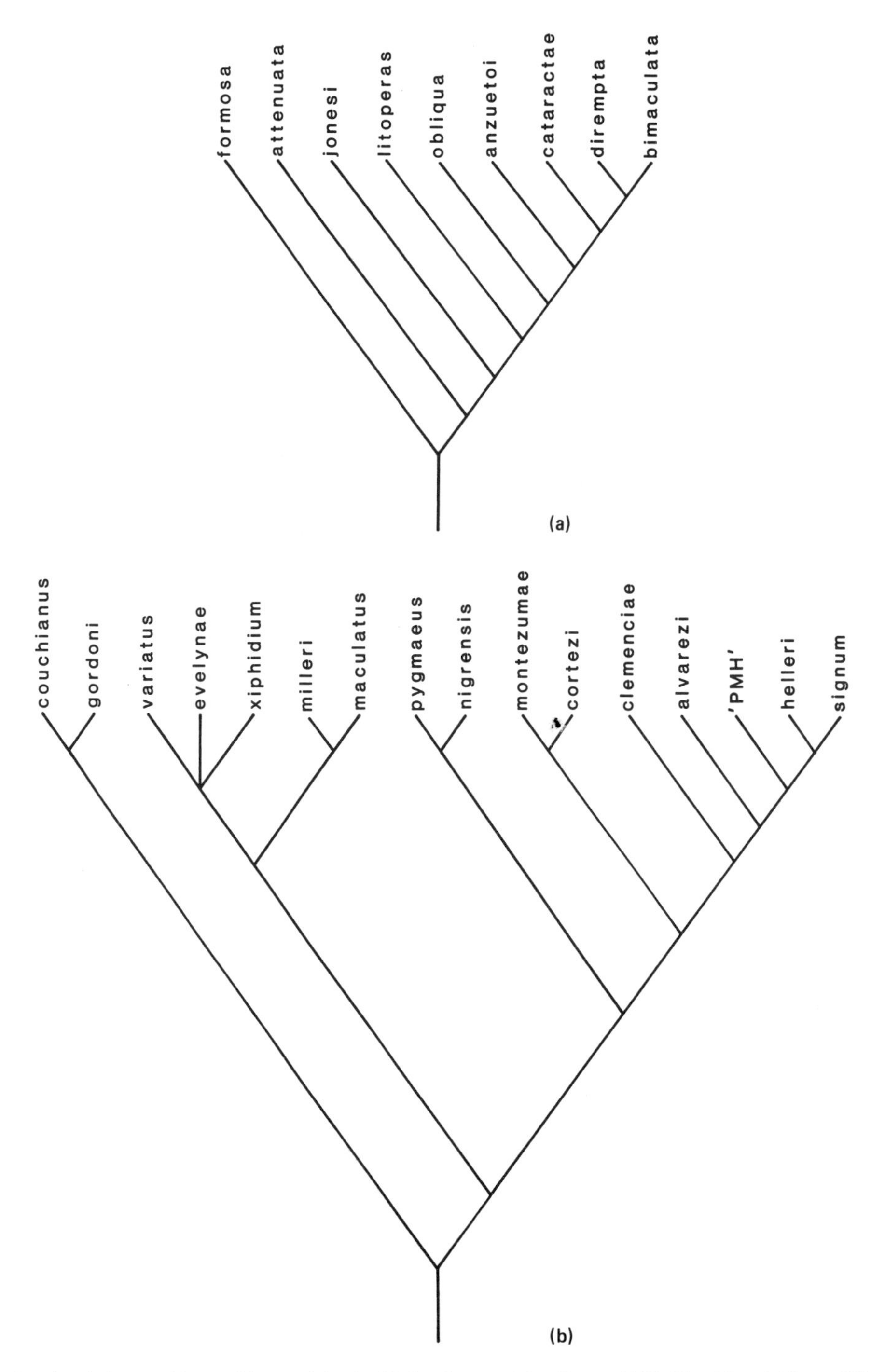

Fig. 2.13. Species cladograms for (a) *Heterandria*, (b) *Xiphophorus*. (After Rosen 1979, Figs 48 and 49, pp. 371 and 372.)

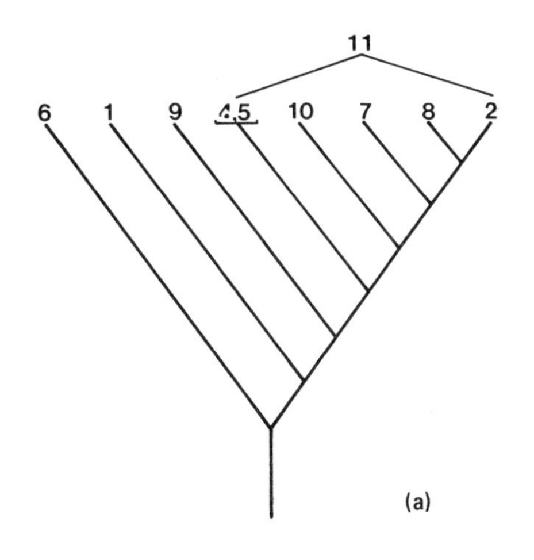

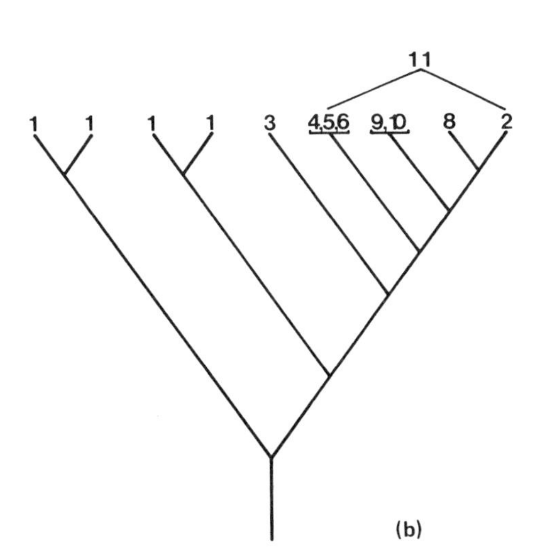

Fig. 2.14. Area cladograms for (a) *Heterandria*, (b) *Xiphophorus*. (After Rosen 1979, Figs 48 and 49, pp. 371 and 372.)

unresolved sequences cannot at the same time be incongruent. Two solutions to this problem have been suggested which are described below.

2.3.3.2 Ancestral species maps—Wiley's method (1980, 1981)

In reviewing Rosen's *Xiphophorus* and *Heterandria* data, Wiley (1980, 1981) took the view that all the speciation events for the Middle American monophyletic groups were the result of successive vicariance events. All of the dichotomies in the cladograms are equivalent to natural barriers arising followed by speciation in both fish genera. Therefore the generalized track of both groups was equivalent to two ancestral taxa. Although this introduced interpretations not necessarily supported by the data Wiley considered the assumption valid since the overall distribution of both groups extended well beyond the Middle American region. This assumption overcame the problem of unique events within each

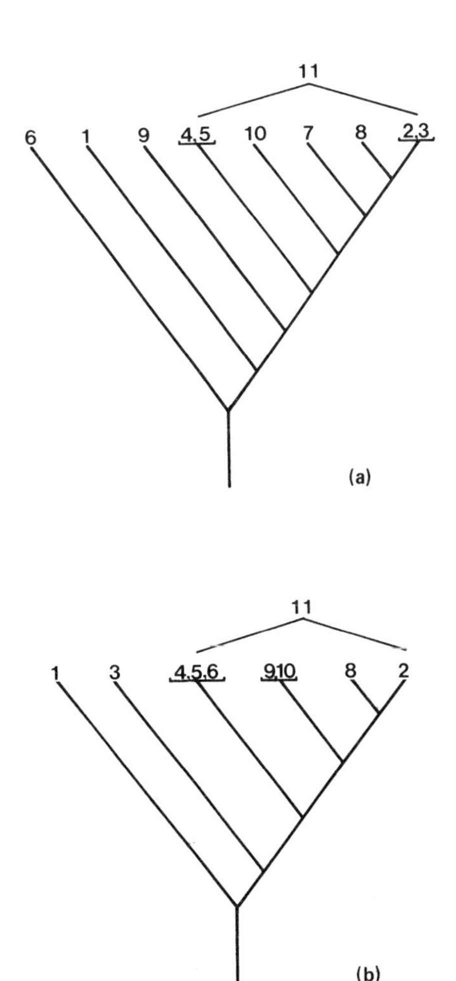

Fig. 2.15. Simplified area cladograms for (a) *Heterandria* and (b) *Xiphophorus* including area components only once. (After Rosen 1979, Figs 48 and 49, pp. 371 and 372.)

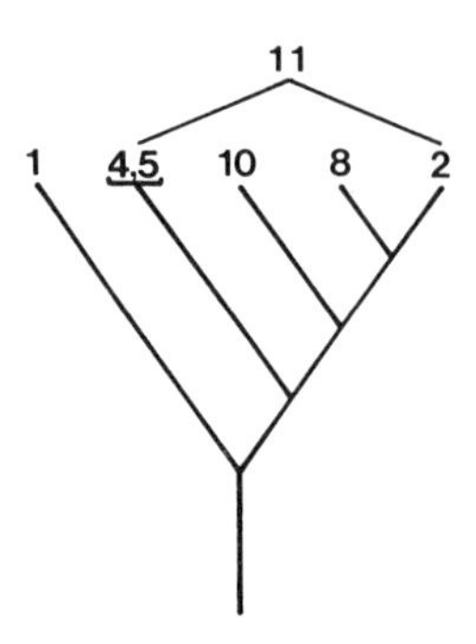

Fig. 2.16. Reduced area cladogram showing area components common to both *Heterandria* and *Xiphophorus*. (From Rosen 1979, Fig. 50, p. 373).

taxon (see Fig. 2.17). Indeed, the very first speciation events were unique for both *Xiphophorus* and *Heterandria*. The initial event was the origin of *Xiphophorus pygmaeus* and *X. nigrensis*, and all other swordtails, from the swordtail common ancestor, and a unique event which isolated *H. attenuata* in area 6 (Fig. 2.17(b)). The second event was common to both genera and involved the vicariance of the whole of area 1 from the rest of the ancestral range. This resulted in the origin of *H. jonesi* and the ancestor of the two remaining area 1 *Xiphophorus* species (*X. montezumae* and *X. cortezi* (Fig. 2.11), although Wiley did not include it on his maps. The subsequent division of these two species further subdivided area 1. It should be concluded that the vicariance event dividing *Xiphophorus* into the two areas 1a and 1b did not affect *Heterandria* at all. Wiley believed two separate and unique events occurred in each genus. *Xiphophorus clemenciae* originated in area 3 and *Heterandria litoperas* originated in area 9 (Fig. 2.17(d)). The next event was common to both groups and involved a separation of a central western part of the ancestral range (Fig. 2.17(e)). The next event separated the southern portion of the remaining ancestral range and the origin of the *Xiphophorus* 'PMH' and *Heterandria anzuetoi* (Fig. 2.17(f)). The next event was the unique origin of *H. cataractae* in area 7 (Figs 2.17(g), 2.10), and finally occurred the peripheral isolation and origin of *Xiphophorus signum* and *Heterandria dirempta* which were isolated in area 8 from the remaining taxa, *Xiphophorus helleri* and *H. bimaculata*, in area 2. To obtain a single hypothesis for the interrelationships of the areas of endemism the cladogram in Fig. 2.18 summarizes the sequence of events for Wiley's successive speciations from an ancestral species. The result provides a complete hypothesis for all eleven areas and is, in this sense, more complete than Rosen's (1978) analysis. However it must be criticized because the method introduces evolutionary assumptions and a degree of interpretation not supported by the data.

2.3.3.3 Component analysis

As we have seen, incongruence can occur for a host of historical reasons. Different groups of organisms exhibit older or younger patterns than the groups to which they are being compared, or they show redundant information due to extinction in one area and unresolved taxonomic groups. Variation in the different patterns, effectively the same as sampling errors, lead to errors in predicting patterns of area interrelationship.

To deal with the problems of redundant, missing, and ambiguous information in cladograms, it is possible to assess the errors caused with a technique called component analysis (Platnick and Nelson 1981). We now give a theoretical discussion of component analysis and would like to consider four areas of endemism, Australia, New Guinea, southern South America and southern Africa, and the hypothetical distributions of eight different monophyletic groups occurring in those areas of endemism (Table 2.1). The first and second groups of lizards and frogs each have one endemic species in each of the four areas of endemism. The third, fourth, and fifth groups—birds, worms, and moths—have different distribution patterns since endemics of each group occur only in three of the four areas. The sixth, seventh, and eighth groups, (trees, ferns, and fish) have three species in each group, one of which is widespread, occurring in two of the four areas. We now discuss each group in turn to demonstrate, that although

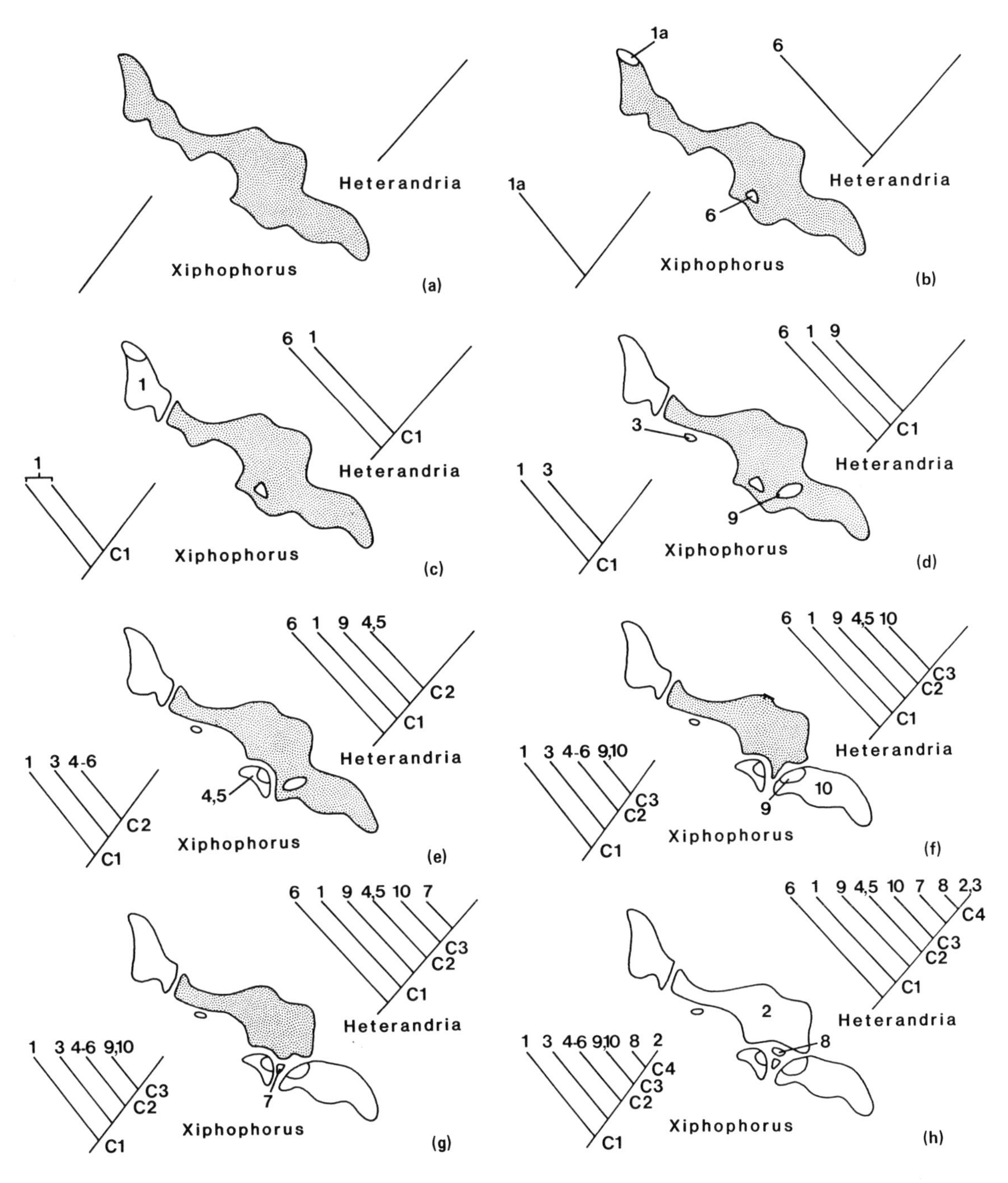

Fig. 2.17. Inferred sequence of speciation and ancestral area maps for *Heterandria* and *Xiphophorus*.

In each diagram stippled areas are inferred ancestral ranges. The phylogenetic position of this ancestral species is represented by the unnumbered branch in the area cladograms above and below the geographic map. Numbered branches correspond to species represented by the number of the area they inhabit. Common speciation events in the history of the two groups are labelled on the area cladograms as C_1, C_2 etc. These labels correspond to the original area cladogram of each species group. (Redrawn from Wiley 1981, Fig. 8.16, p. 303.)

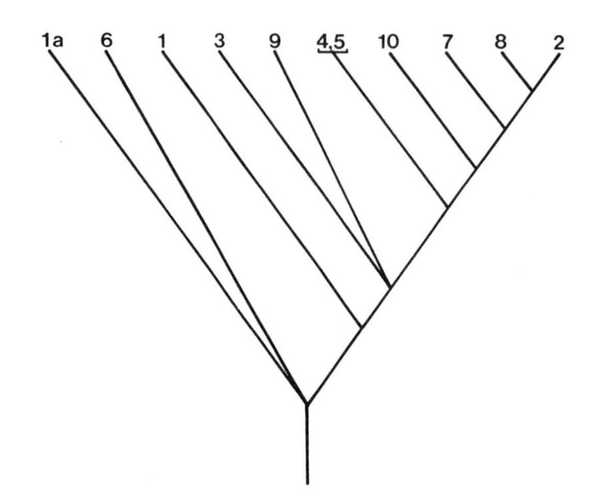

Fig. 2.18. Consensus cladogram for *Heterandria* and *Xiphophorus* based on Wiley's ancestral area maps. See Fig. 2.17 and text for explanation.

each group has a unique pattern of area relationships, each pattern has the same biogeographical history.

2.3.3.4 Endemics

The cladograms of the frogs and lizards are identical. The endemics of Australia and New Guinea are more closely related to one another than either are to those of South America and Africa. The South American taxa are more closely related to these western Pacific taxa than to the African ones. The general relationships of each group are expressed as area components denoted by numbers on a combined cladogram (Fig. 2.19). This cladogram shows a pattern remarkably congruent to the geological cladogram (Fig. 2.20) indicating the relative histories of the four southern hemisphere areas (based on Rosen 1978). One conclusion is that the lizard and frog biota and the earth have the same historical sequence since they all yield the same components 1, 2, and 3.

2.3.3.5 Missing areas

The birds, worms, and moths (Fig. 2.21) are characterized by having three endemic taxa in

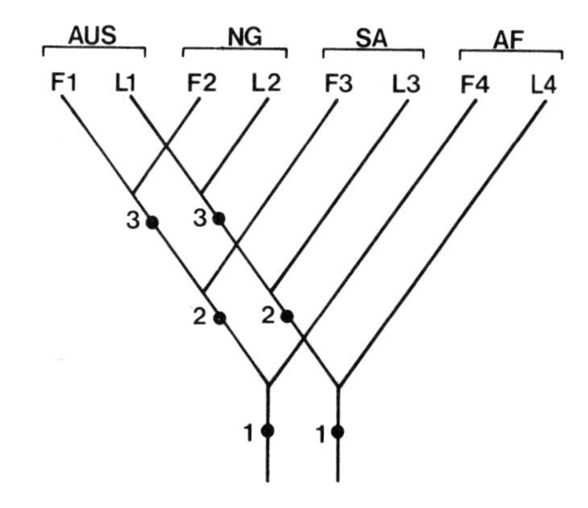

Fig. 2.19. Hypothetical cladograms for frogs (F) and lizards (L) occurring in Australia (AUS), New Guinea (NG), South America (SA), and Africa (Af). The identical area components are shown as dots labelled 1, 2, and 3.

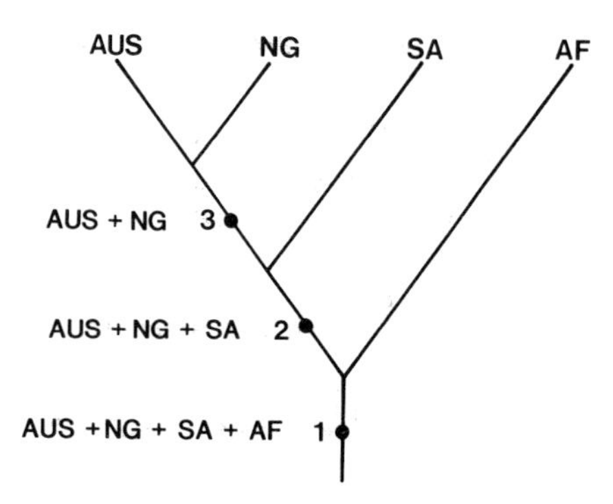

Fig. 2.20. Geological area cladogram with the area components for the four areas shown as dots and numbered 1, 2, and 3.

Table 2.1

Distribution of eight hypothetical taxonomic groups in the southern hemisphere (see text for explanation; see also Figs 2.19-2.27)

Frogs	(F)	AUS;	NG;	SA;	AF
Lizards	(L)	AUS;	NG;	SA;	AF
Birds	(B)	AUS;	NG;	—;	AF
Worms	(W)	AUS;	—;	SA;	AF
Moths	(M)	AUS;	NG;	SA;	—
Trees	(T)	AUS;	NG;	*SA/AF*	
Ferns	(Fe)	*AUS/NG*;		SA;	AF
Fish	(Fi)	*AUS/AF*;	NG;	SA	

Key to areas: AUS, Australia; NG, New Guinea; SA, South America; AF, South Africa.

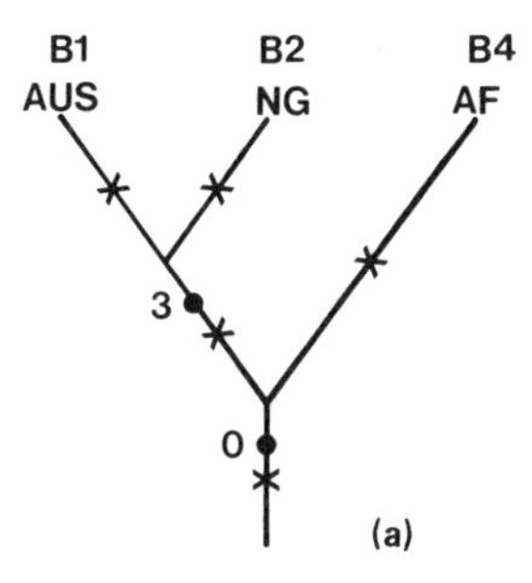

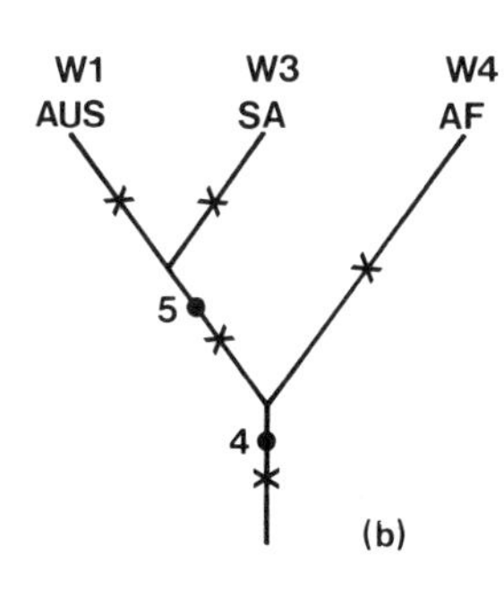

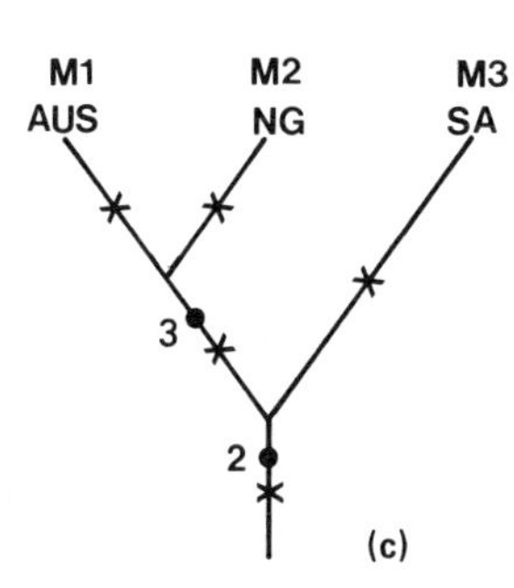

Fig. 2.21. Hypothetical cladograms for three groups: (a) birds; (b) worms; and (c) moths, with endemics occurring in three of the four areas. The area components are shown as dots and numbered. For explanation of Xs see text.

three of the four areas of endemism. Examining the species interrelationships and substituting the taxa for areas gives us three more area components than those from the lizards and frogs for Australia, New Guinea and Africa (0) (Fig. 2.21(a)), Australia and South America (5) (Fig. 2.21(b)), and Australia, South America, and Africa (4) (Fig. 2.21(b)). Since each of the area cladograms has only three areas, each has only two rather than three of the components necessary to give a cladogram of the four areas. In other words, they all lack the four-area component 1, and if the three different cladograms represent a pattern due to similar histories then some of the apparent two- and three-area components must be false. As each cladogram has three of the four areas, each can be restored to a four-area cladogram by adding the missing area. The missing areas can be inserted in one of five positions as indicated by the crosses in Fig. 2.21. For example, consider the bird cladogram in Fig. 2.21(a). By adding the missing South American area the five fully restored four-area cladograms can be seen in Fig. 2.22(a)-(e). Similarly, for the worm cladogram in Fig. 2.21(b), the missing New Guinea area can be found in the five four-area cladograms of Fig. 2.22(a, c, d, g, i). Finally, the missing African area in the moth cladogram of Fig. 2.19(c) can be found in the five four-area cladograms of Fig. 2.22(a, b, e, f, h). Nine of all the fifteen possible cladograms are thus reiterated. Four of the cladograms (Fig. 2.22(f, g, h, i)) agree with only one of the three original three-area bird, worm or moth cladograms in Fig. 2.21; four more (Fig. 2.22(b)-(e)) agree with only two of the three original three area cladograms, and only one (Fig. 2.22(a)) agrees with all three of the three-area cladograms. Thus, if the patterns of the birds, worms, and moths are due to the same historic events then the simplest hypothesis is the one which agrees with the original geological cladogram (Fig. 2.22(a)). This hypothesis recognizes components 1, 2, and 3 as real and components 0, 4, and 5 as false due to the missing taxa. By taking a naïve approach and considering all the components 0–5 as real, a problem emerges when assessing the information content of the conflicting statements. The effect of this problem can be seen in Fig. 2.23(a)-(c). By adding the bird and worm cladograms together in Fig. 2.21(a)-(b), the four-area component 1 is recovered but there is no way of resolving components 3 and 5 since either is equally plausible and thus Australia, New Guinea, and South America (component 2) must be left as

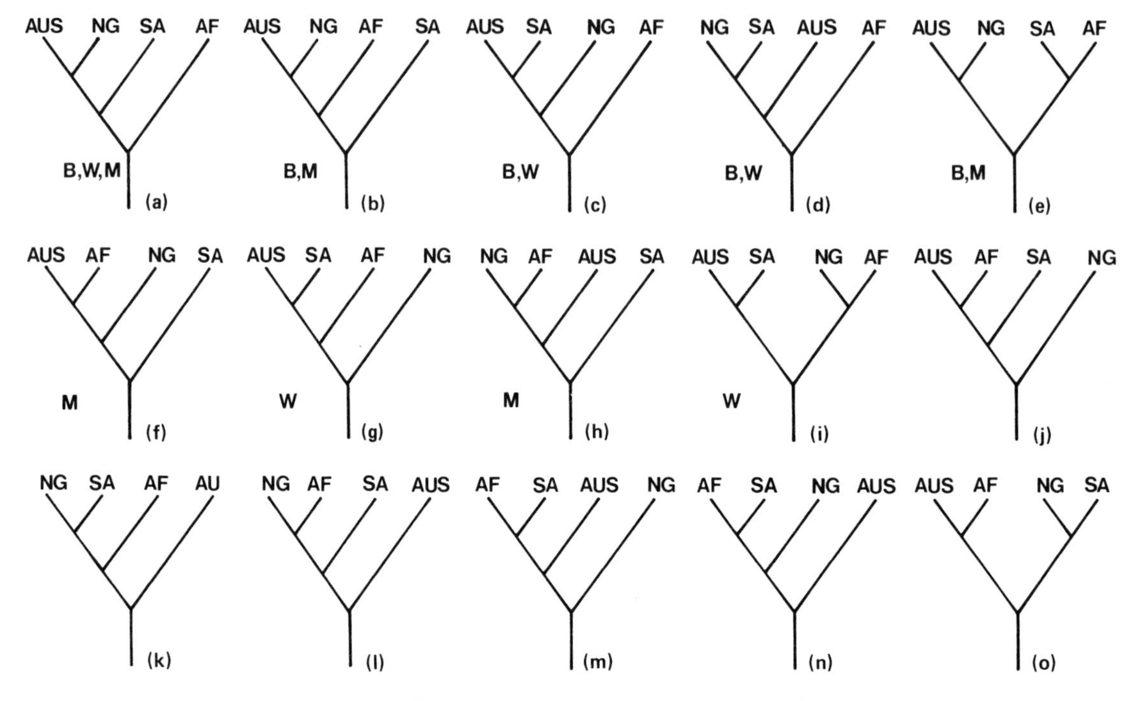

Fig. 2.22. Fifteen possible area cladograms for four areas. See text for explanation.

an unresolved trichotomy. Similarly, by adding the bird and moth cladograms of Fig. 2.21(a), (c) together, the Australia and New Guinea component 3 is common to both and the four-area component 1 is recovered; but the incongruence for the 0 and 2 components leaves the relationships of South America and Africa with the other two areas unresolved. A totally unresolved pattern occurs when all three three-area bird, worm, and moth cladograms in Fig. 2.21 are added together (Fig. 2.23(c)).

2.3.3.6 Widespread taxa

The sixth, seventh, and eighth groups of hypothetical taxa, the trees, ferns, and fish, exhibit different kinds of distributions whereby one species in each group of three has a widespread distribution occurring in two of the four areas (Fig. 2.24). A comparison of these area cladograms indicates that each is lacking in some information because they can be considered to have unresolved area interrelationships and incongruent distribution patterns (e.g. Fig. 2.24(a), (b), (c)).

As was seen with the birds, worms, and moths, with taxa missing from certain areas the patterns can be considered under one of two possible assumptions; either all of the identifiable area components are meaningful or we can take a less than naïve approach and analyse for missing information by admitting that some of the components must be false. The full theoretical aspects of analysing under these different assumptions, which we will call assumptions 1 and 2, have been considered by Nelson and Platnick (1981) and Nelson (1982).

Let us consider our three different distribution patterns under both assumptions. Under assumption 1, whatever is true of the widespread tree in one part of its range (e.g. $T_{3,4}$ in South America; Fig. 2.25(c)) must also be true of the same tree species in another part of its range (e.g. $T_{3,4}$ in Africa). However, under assumption 2 whatever

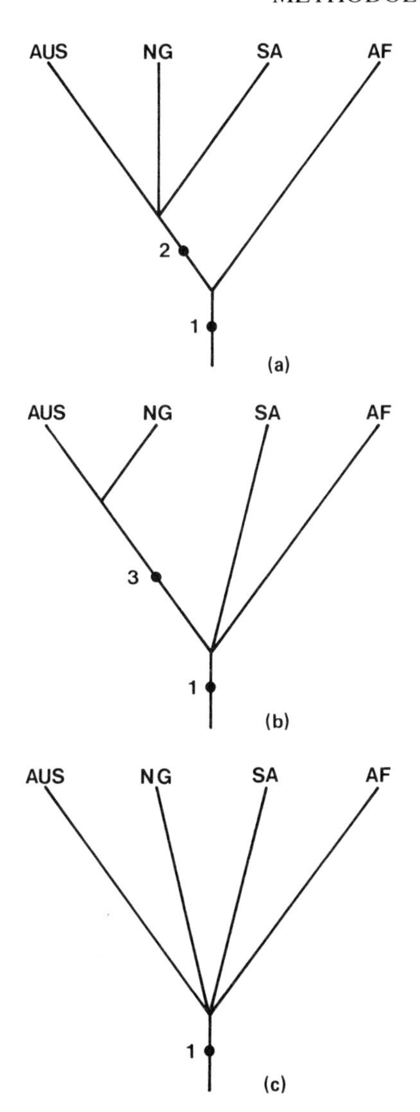

Fig. 2.23. Consensus trees for four areas obtained by adding together the three area cladograms from Fig. 2.21. (a) Birds and worms; (b) birds and moths; (c) birds, worms, and moths. Area components are shown with black dots and numbered.

is true for the tree species ($T_{3,4}$) in one part of its range (in South America) need not also be true of the same tree species elsewhere (in either South America or Africa, but not in both).

The implications under assumption 1 are that the widespread tree species occurring in both South America and Africa will never be distinguished as two separate taxa. In other words, if the tree species T_2 in New Guinea and T_1 in Australia are more closely related to one another than to $T_{3,4}$ in South America, then the Australian and New Guinea species are more closely related to one another than they are to the widespread species ($T_{3,4}$) in Africa.

Under assumption 1 the area cladogram yields only components 1 and 3 (Fig. 2.25(c)). Such a partially resolved cladogram under further analysis would allow for only three possible dichotomous cladograms (Fig. 2.25(d), (e), and (f)).

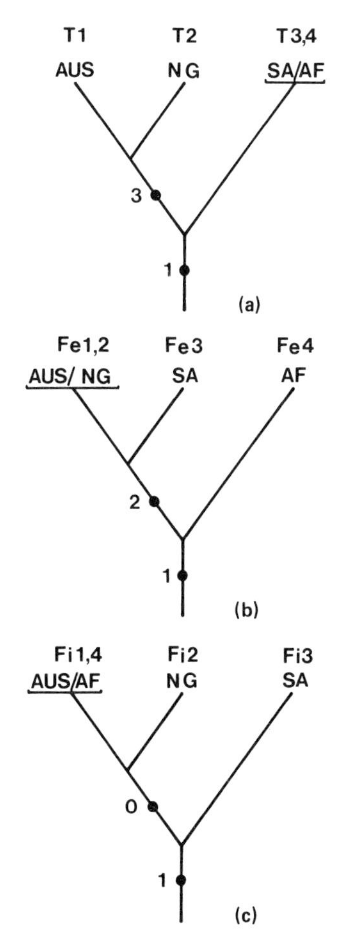

Fig. 2.24. Hypothetical cladograms for three groups of three taxa with one species in each group widespread to two areas. (a) Trees, (b) ferns (c) fish.

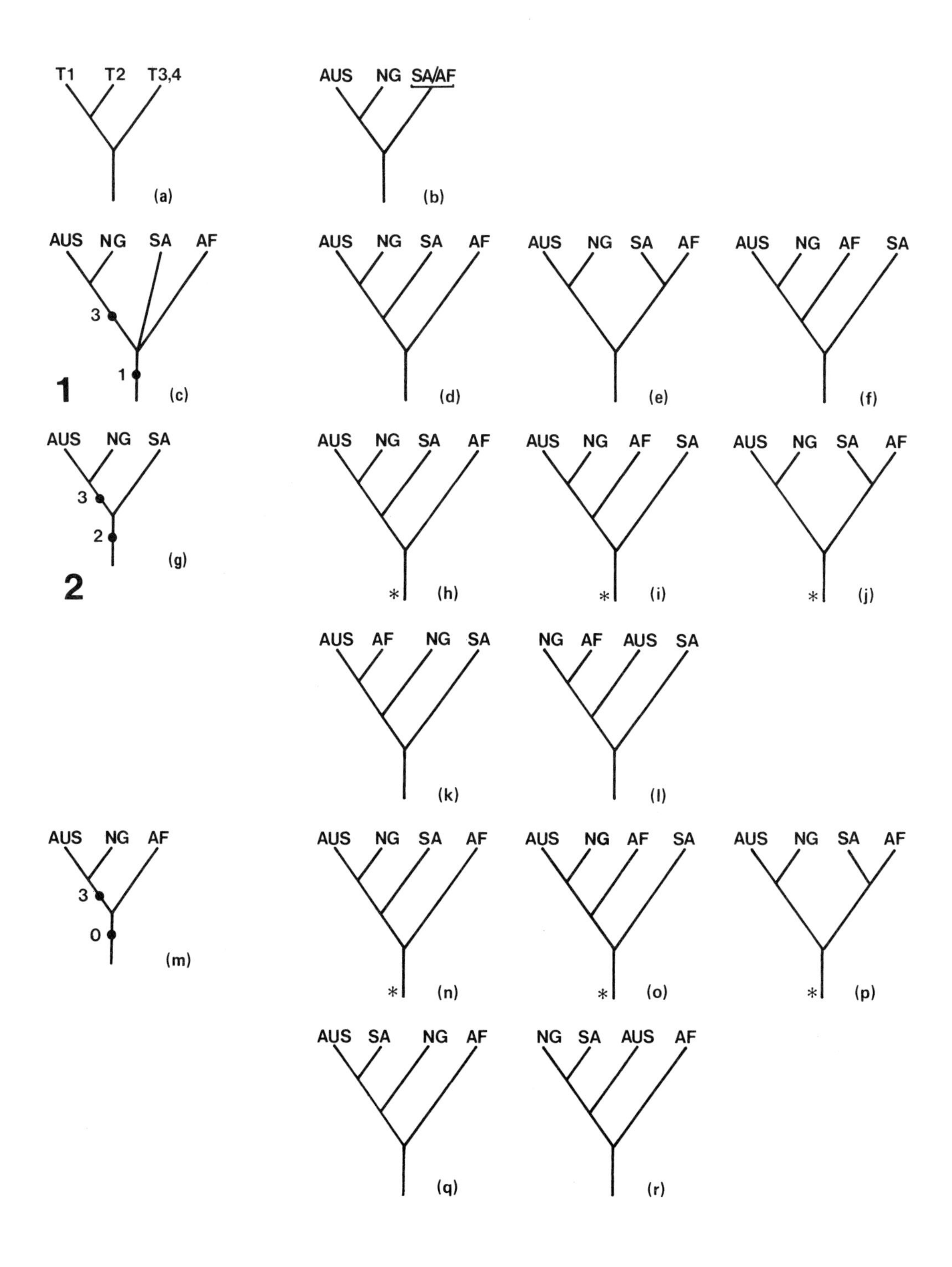

T1 T2 T3,4
(a)
AUS NG SA/AF
(b)
AUS NG SA AF
3
1
1
(c)
AUS NG SA AF
(d)
AUS NG SA AF
(e)
AUS NG AF SA
(f)
AUS NG SA
3
2
(g)
2
AUS NG SA AF
*
(h)
AUS NG AF SA
*
(i)
AUS NG SA AF
*
(j)
AUS AF NG SA
(k)
NG AF AUS SA
(l)
AUS NG AF
3
0
(m)
AUS NG SA AF
*
(n)
AUS NG AF SA
*
(o)
AUS NG SA AF
*
(p)
AUS SA NG AF
(q)
NG SA AUS AF
(r)

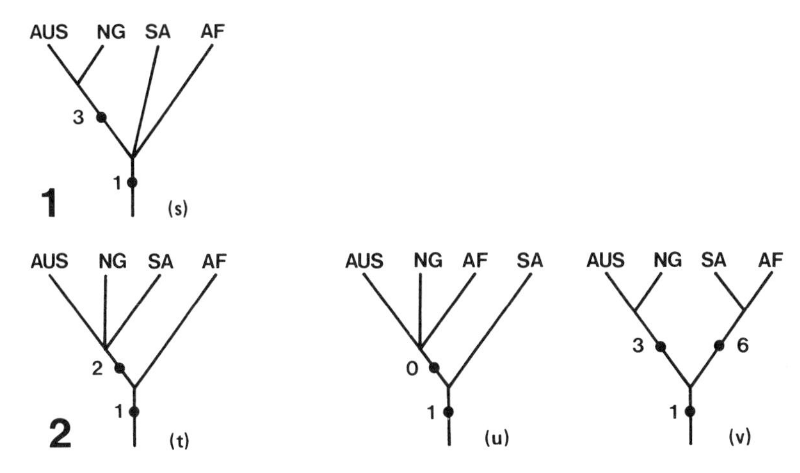

Fig. 2.25. Component analysis for flowering trees. (a) Cladogram for two endemic trees in Australia and New Guinea and one widespread species in South America and Africa. (b) Area cladogram. (c-f) Component analysis under assumption 1. (g-r) Component analysis under assumption 2. (s) Summary components from assumption 1. (t-v) Summary components under assumption 2.

Under assumption 2 the widespread occurrences in South America and Africa of $T_{3,4}$ might at some time be divided into two separate entities, such that whatever is true of one occurrence might not necessarily be true of the other. Using the same example, species T_1 in Australia and species T_2 in New Guinea could be more closely related to each other than either is to $T_{3,4}$ in South America, or in Africa, but not in both. Under this assumption, the four-area cladogram (Fig. 2.25(b)) yields two possibilities for component analysis (Fig. 2.25(g), (m)) but each possibility includes only three of the four areas under analysis. The two components 2 for Australia, New Guinea, South America and 0 for Australia, New Guinea, Africa each allow for five different fully dichotomous cladograms (Fig. 2.25(h)-(l), (n)-(r)) for the four areas. A comparison of the two rows of five cladograms shows that three in each are repeated (Fig. 2.25(h)-(j), (n)-(p), as shown by the asterisks) giving a total of seven possible cladograms that when compared to each other can be reduced to five unique components (Fig. 2.25(t)-(v)). If the Australia/New Guinea component (3) is correct then either the Australia/New Guinea/South America component (2) or the South America/Africa component (6) can be correct. If the South America/Africa component (6) is correct then only the Australia/New Guinea component (3) can be correct. However, if the Australia/New Guinea/South America component (2) is correct then there are three possible two-area components for Australia/New Guinea, Australia/South America or New Guinea/South America. This conflicts with yet another three-area component, Australia/New Guinea/Africa (0) which allows for three possible two-area components, Australia/Africa, New Guinea/Africa and Australia/New Guinea, and is so left as a trichotomy. Thus, the problem of one widespread species is that it gives ambiguous results which will only have value if compared with other widespread groups.

Similar analyses have been carried out for the two other cladograms with widespread species in each case occurring in two different areas (Figs 2.26, 2.27). Under assumption 1 there are three possible cladograms which when combined yield two components for each group (Figs 2.26(a) and (b) and 2.27 (a) and (b), components 0, 1 and 1, 2 respectively). Under assumption 2 (Figs 2.26(b)-(f) and 2.27(b)-(f)) there are seven possible different cladograms which give eight equally likely components in each case.

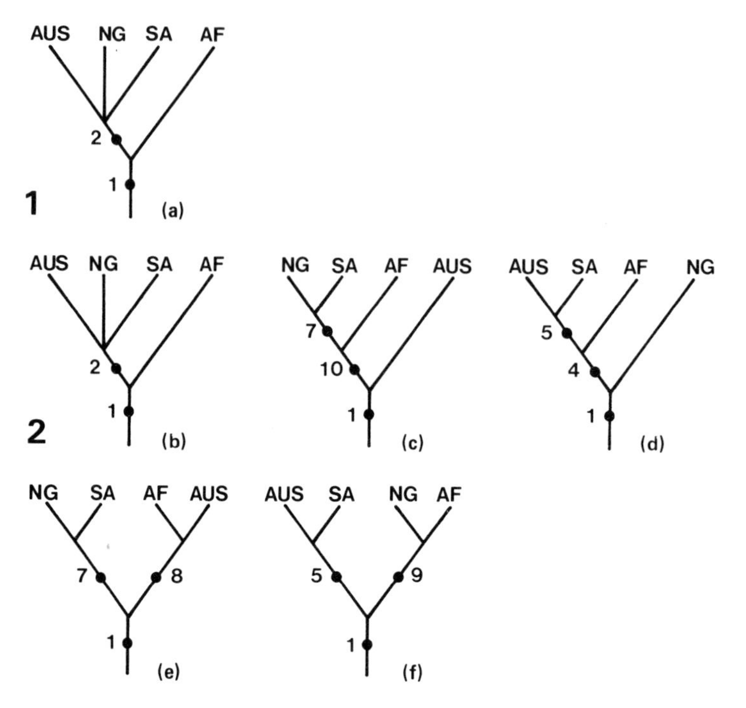

Fig. 2.26. Summary components for ferns (see Fig. 2.24) with endemic taxa in South America and Africa and a widespread species in Australia and New Guinea. (a) Summary components under assumption 1. (b-f) Summary components under assumption 2.

The component analyses emphasize the problem of not having a way of determining which hypotheses to choose when considering only one group of taxa. The interesting results emerge by combining the information. Combining the different area cladograms (Figs 2.28 and 2.29) is equivalent to combining the components of each group to find a resolution. Consider the possibility of combining two cladograms, with widespread trees in South America/Africa and a widespread fish species in Australia/Africa (Fig. 2.24(a), (c)). If added together under assumption 1 the two conflicting components 0 and 2 would result in a partially uninformative consensus diagram (Fig. 2.28(c)-(e)). Under assumption 2 there are several possibilities of which just two are shown (Fig. 2.28(f)-(h), (i)-(k)). For example, take the implied cladogram in Fig. 2.28(f) (=Fig. 2.25(t)). The only implied cladogram with which it can be combined to give an informative result is that of Fig. 2.28(g) (=Fig. 2.27(c)). However, a different result is obtained for the implied cladograms shown in Fig. 2.28(i) (=Fig. 2.25(u)) which can only be combined with Fig. 2.28(j) (=Fig. 2.27(d)). These results mean that when only a few groups are available for analysis assumption 2 will give resolved results by comparison to assumption 1 but more than one final hypothesis is obtained. However, when all three, tree, fern, and fish area cladograms containing widespread taxa are combined (Fig. 2.29(a)-(c)) only one fully informative consensus cladogram can be produced (Fig. 2.29(k); from the implied cladograms in Figs 2.25(t), 2.26(b), 2.27(c)) under assumption 2. When combining them under assumption 1, the result is partially uninformative (Fig. 2.29(d)-(g)).

Sometimes, combining the implied cladograms for the trees and ferns under assumption 1 can be totally informative (Fig. 2.29(l)-(n)). However,

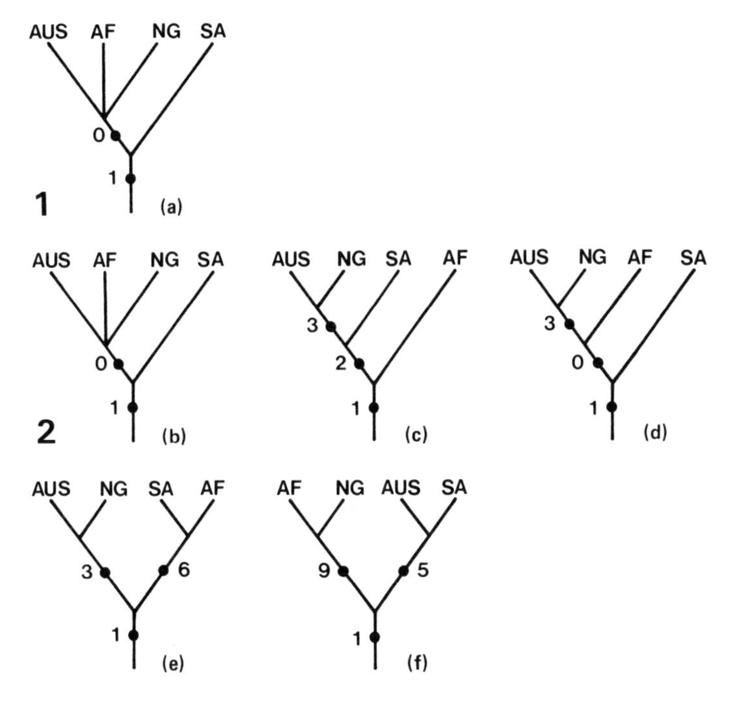

Fig. 2.27. Summary components for fish (see Fig. 2.24) with endemic taxa in South America and New Guinea and a widespread taxon in Australia and Africa. (a) Summary components under assumption 1. (b-f) Summary components under assumption 2.

this happens only when there is no overlap or conflict in the original area cladogram. Assumption 2 is far less restrictive than assumption 1. Generally if there is information on area interrelationships that can be extracted from ambiguous data such as that found with widespread taxa then it is most likely to be obtained with assumption 2. A comparison of the best hypothesis (Fig. 2.29(k), (n)) with the original geological cladogram corroborates the general hypothesis for the history of the four areas (Fig. 2.20).

2.3.3.7 Platnick's example (1981)

The two quite widely distributed poeciliid fish genera *Heterandria* and *Xiphophorus* (Figs 2.10, 2.11) each has 11 identifiable disjunct areas (Fig. 2.12). The area cladograms are shown in Fig. 2.30. Areas 4 and 5 are occupied by one species in each genus and are thus treated as a single area. Area 11 was treated by Rosen (1978, 1979) as a putatively hybrid area between areas 4, 5 and 2 but since it is in fact a true disjunct area, it is maintained here. The area 11, containing hybrid information, can be treated as unresolved and placed at its ancestral stem (Fig. 2.30).

A comparison of the two cladograms (Fig. 2.30(a), (b)) shows that *Xiphophorus* is less informative than *Heterandria* because it has two widespread species in areas 4, 5, 6, 9, and 10 and is absent from area 7. In *Heterandria* areas 4, 5, 6, 9, and 10 are all occupied by recognizable endemic taxa.

As we saw in the last section, under assumption 1 whatever is true of a widespread taxon in one part of its range (e.g. *Xiphophorus alvarezi* in area 4, 5) must also be true in the other part of its range (i.e. area 6). However, under assumption 2 whatever is true of a widespread taxon in one part of its range need not also be true of the taxon elsewhere. In other words, the widespread distributions are

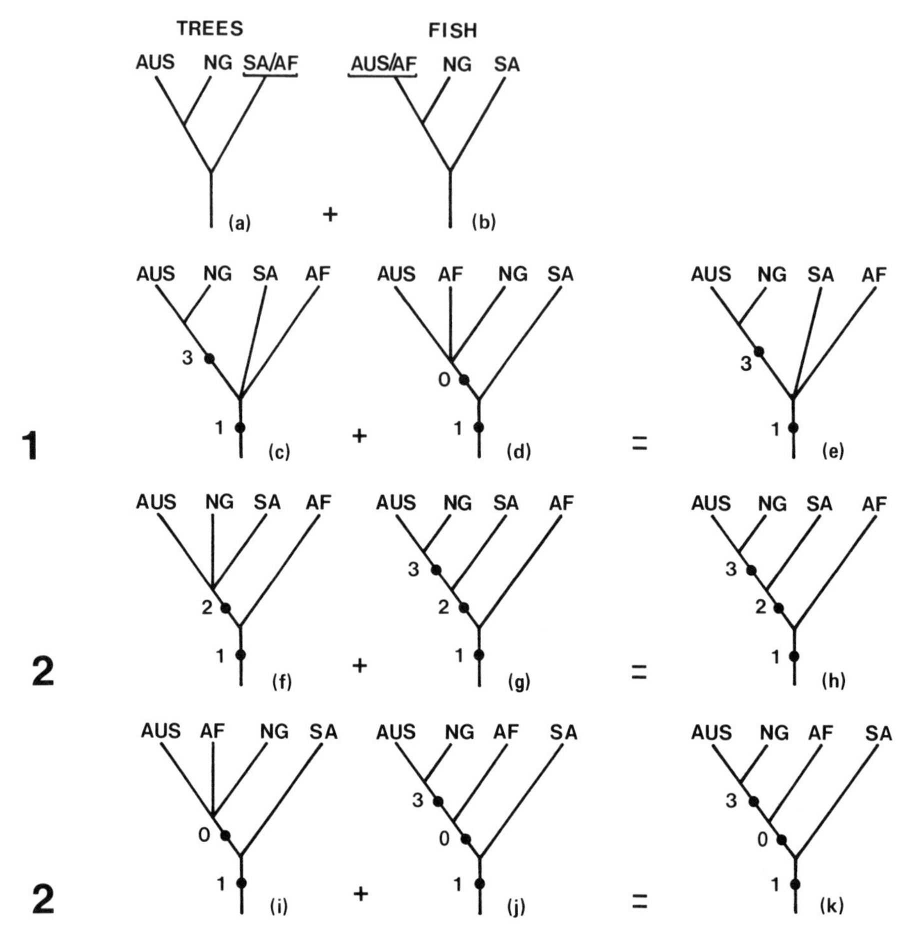

Fig. 2.28. Consensus diagrams for trees and fish. (a, b) Area cladograms. (c)-(e) Combining summary components under assumption 1. (f)-(h), (i)-(k) Two possible solutions arrived at by combining summary components under assumption 2.

equivalent to saying that we are ignorant of the reasons for lack of resolution in the cladogram. In distributional terms, that is equivalent to saying that we do not know whether the patterns are due to dispersal or a failure to speciate in response to a vicariance event. Rosen's original application of the Platnick and Nelson (1978) biogeographic method in 1978 and 1979 compared the two cladograms to one another and identified only those parts which were congruent, which meant that a cladogram for only six areas could be produced (Fig. 2.31(a)). Fig. 30 shows two area cladograms for *Heterandria* and *Xiphophorus* corrected for area 11. Platnick (1981) considered the removal of unique and incongruent areas as equivalent to analysing under assumption 1. If assumption 1 is adopted, then the *Xiphophorus* populations of area 9 must be most closely related to the population in area 10, and the information for area 9 is incongruent with the information from *Heterandria*. Similarly, the information on area 6 is also incongruent to both cladograms.

A different result can be obtained by applying assumption 2. By taking the information on areas

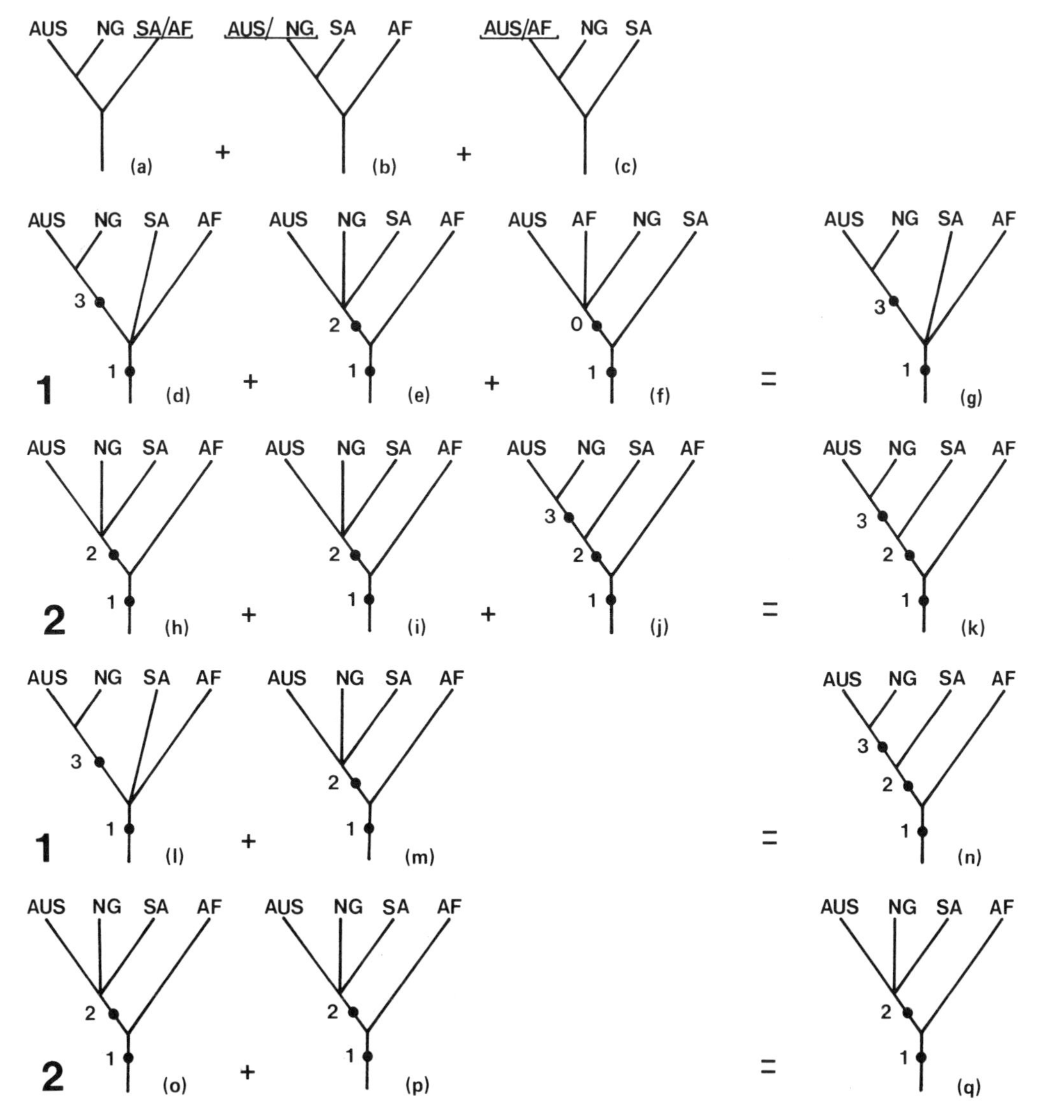

Fig. 2.29. Consensus diagrams for trees, ferns and fish. (a), (b), (c) Area cladograms for each group. (d)–(g) Combining summary components under assumption 1. (h)–(k) One solution by combining only possible summary components under assumption 2. (l)–(n) Combining trees and ferns under assumption 1; (o)–(q) combining trees and ferns under assumption 2.

6 and 9 from *Heterandria* as correct then the incongruent information in the same areas for *Xiphophorus* is due either to dispersal or a failure to speciate in response to a vicariance event. Rosen's original reasons for applying a version of assumption 1 was that when groups dispersed or failed to respond to vicariance events it reduced informativeness. Platnick (1981) noted however, that if widespread taxa are uninformative they cannot be incongruent (as well). Absence data can never be incongruent with information at hand so unique areas should never be deleted on these

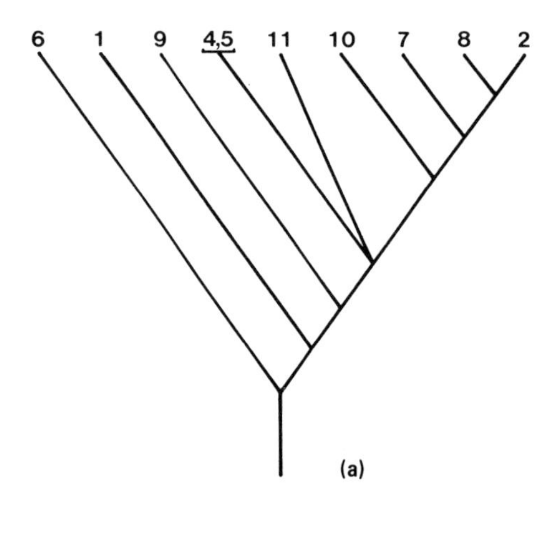

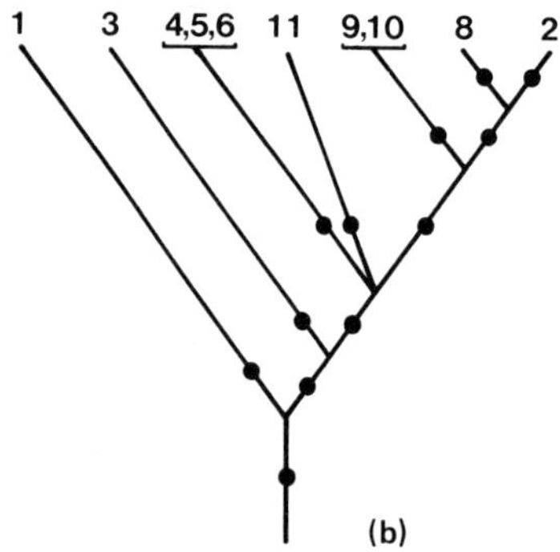

Fig. 2.30. Area cladograms for (a) *Heterandria* and (b) *Xiphophorus* corrected for area 11. For explanations see text.

We have a cladistic structure that accounts for all 11 areas of endemism that can be recognized from the two fish genera *Heterandria* and *Xiphophorus*. We emphasized that this gives an almost identical result for the 11 areas to that obtained by Wiley's method but was obtained by using cladogram logic rather than evolutionary assumptions. If such a pattern is due to changes in earth history the question that we could ask now is what might have been the historical factors in Mesoamerica to cause this pattern and how might these be compared with the given biological distribution? So that biotic and historical patterns can be compared we would ideally require that geological grounds. Taken on their own, widespread taxa under assumption 2 give uninformative components but when considered with other cladograms involving widespread taxa informative results are possible. Under assumption 2 the *Xiphophorus* cladogram (Fig. 2.30(b)) allows the populations in area 9 or 10 (but not both) and area 4, 5, or 6 (but not all) to occur in any of twelve positions, on any line in the cladogram (as shown by the black dots). The analysis yields three possible cladograms all of which are plausible. These can be summarized in one cladogram by a trichotomy as shown by an asterisk in Fig. 2.31(b). A sequence of vicariance events implied by this cladogram is illustrated in Fig. 2.32.

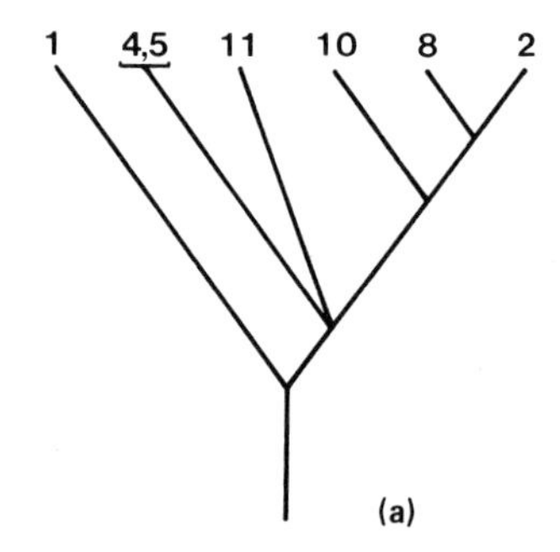

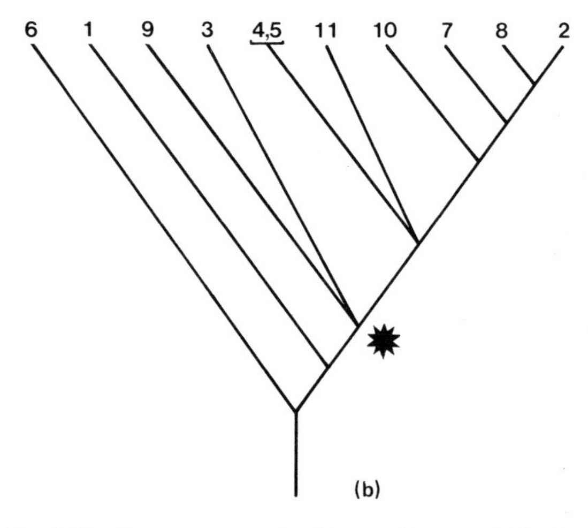

Fig. 2.31. Consensus trees for *Heterandria* and *Xiphophorus* analysed under (a) assumption 1; (b) assumption 2.

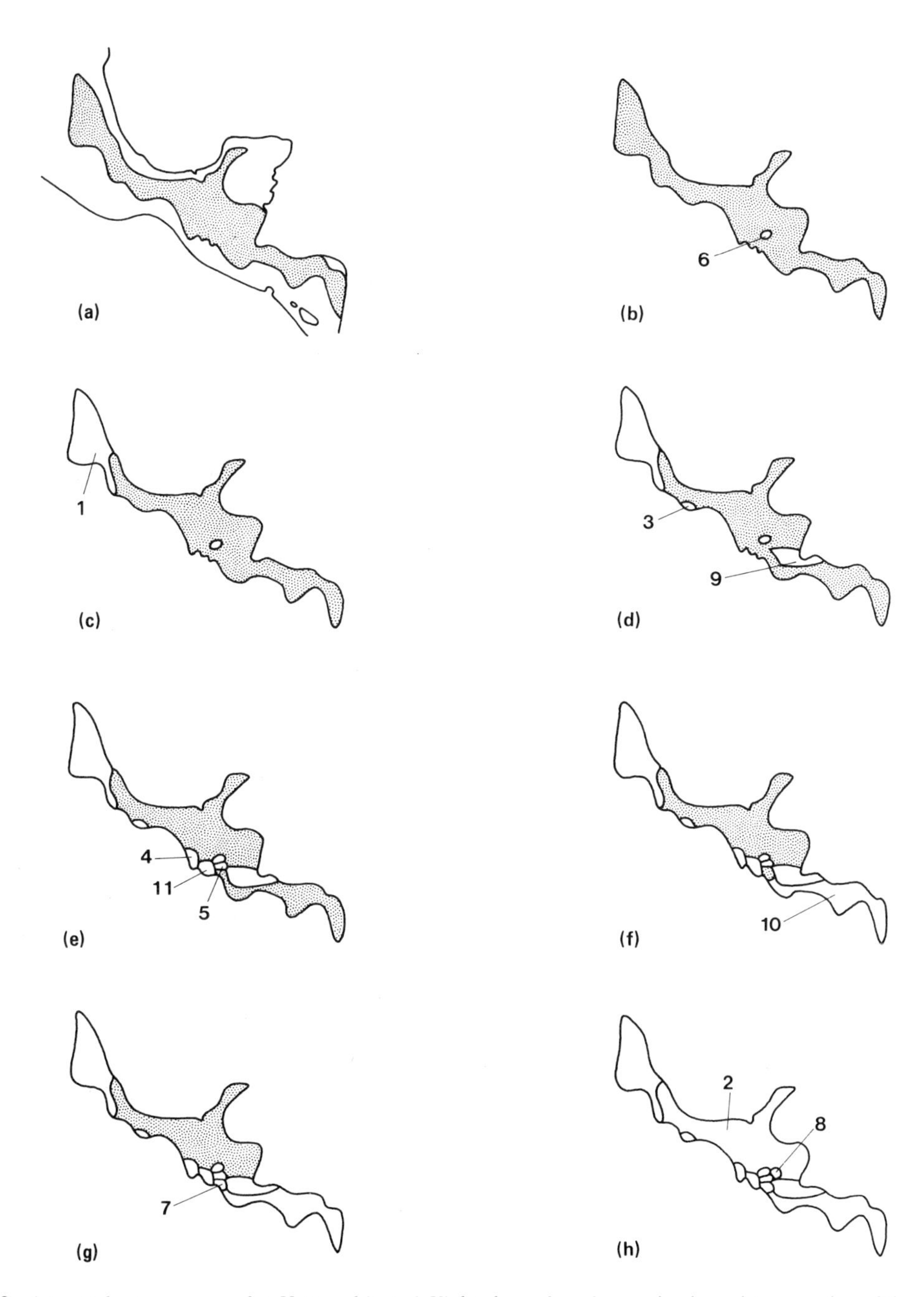

Fig. 2.32. Ancestral sequence map for *Heterandria* and *Xiphophorus* based on a vicariance interpretation of the consensus tree in Fig. 2.31 (b). See text for explanation.

information be assembled into cladograms in the same way as biological cladograms. Until such time as geological data can be ordered for a more informative comparison one can say little except that the observed patterns in Mesoamerica have been formed over a period of at least the last 80 million years (Rosen 1978).

2.4 CONCLUSIONS

We hope that, by outlining the methods of cladistic biogeography and describing in detail Rosen's pioneering work on Central American poeciliid fish, we have underlined the significance of cladistics for historical biogeography. Comparing Rosen's results with Good's (1974) relatively traditional approach and Croizat's (1958) panbiogeographic method emphasizes the point further. On the basis of plant distributions Good indicated that 'Isthmian America . . . has a close relationship with western North America' (Good 1974, p. 156). He also said:

> There is a well-defined line of demarkation (usually described as running from the Gulf of Fonseca on the Pacific coast southeast across Nicaragua to the Caribbean) which may be regarded as the southern limit of the isthmian extension southward of the Rocky mountain system. This area is geologically much younger and has undergone various degrees of subsidence and emergence since the Cretaceous until, in the Oligocene, a great upheaval resulted in the formation of the lands which, to quote a vivid phrase 'finally sealed the marine portal between North and South America. (Good 1974, p. 233.)

In other words, only one two-area biogeographic statement is expressed, isthmian America is most related to western North America! Croizat's (1958) work is far more detailed. He devotes the whole of chapter VII of *Panbiogeography* Vol. 1 to Central American and Caribbean biogeography and describes a variety of botanical and zoological examples. The main conclusions are best summarized by Rosen (1978) as we noted on p. 29 (see Fig. 2.7). Track analysis yields five main components; a North American-Caribbean track, a South American-Caribbean track, an eastern Pacific-Caribbean track, an eastern Atlantic-western Atlantic track and an eastern Pacific-eastern Atlantic track. In effect, the result is a series of five general similarity statements. Cladistic biogeography goes much further. As we have seen from Rosen's poeciliid fish data and using the method of Nelson and Platnick (1981) it is possible to resolve eleven areas of endemism within Central America into a single hierarchy.

It seems then at this time that methods of cladistic biogeography are the best to analyse and compare biotic patterns at the highest resolution so as to compare them to independent sources of data such as geological patterns. Furthermore, cladistics is a general method of determining class and subclass relations whatever the source of data without recourse to evolutionary narrative (see Nelson 1982). A cladistic view of world history combined with the cladistic method in systematics makes it possible to express area interrelationships as hierarchical relations from biotic information. The development of methods, such as Nelson and Platnick's assumption 2, allows the possibility of generating general biogeographic hypotheses from congruent cladograms even from seemingly ambiguous patterns. Assumption 2 is a general empirical procedure without any dispersal, vicariant or extinction events assumed in the analysis but which at the same time never denies that they occur (Nelson 1982).

3 THE REAL WORLD

3.1 INTRODUCTION

We have established as one of the aims of cladistic biogeography the production of area cladograms—hierarchical relationships of areas derived directly from cladograms of taxa. Inherent in our support for this aim is the principle that biogeographic studies must be based on sound systematics, a notion nearly universally upheld by the biogeographers writing before us (e.g. Brundin 1966; Croizat 1964; Darlington 1957, 1965; Good 1964; Rosen 1976). Where vicariance and cladistic biogeographers diverge most significantly from dispersalist biogeographers is in the idea that schemes of relationship of taxa (cladograms) within a biota are the basic and fundamental characteristics of any biogeographic study, rather than, for example, hypotheses of dispersal capabilities of individual taxa.

Examination of cladograms of different taxa from the same 'areas of endemism' for the same area relationships constitutes a search for a biogeographic pattern. These patterns are what we claim need to be explained in historical biogeographic studies. For example, a closer relationship of a taxon in area A to a taxon in area B, than to one in area C (Fig. 3.1(a)) in nine independent cases is a biogeographic pattern. The pattern specifies a corroborated set of relationships among areas A, B, and C. We may look for an explanation of the pattern in the geological histories of areas A, B, and C, such as the separation of land mass C from A and B before the separation of taxa in A and B. A single exception to this pattern (Fig. 3.1(b)) in which a taxon in area B is more closely related to a taxon in area C than to a taxon in area A is an alternative pattern, yet does not necessarily demand a general explanation. By definition, a general explanation is applied to more than one phenomenon or in this case more than one cladogram of areas. The pattern of Fig. 3.1(b) may be explained as an individual case by random dispersal or perhaps a sampling error which could always be responsible for a scheme of relationships in contrast to the general pattern. Such an explanation would be unique to the taxon, not to the biota unless random dispersal were the rule. The conflicting pattern might also mean that area B is a hybrid area and should be treated as two components rather than one.

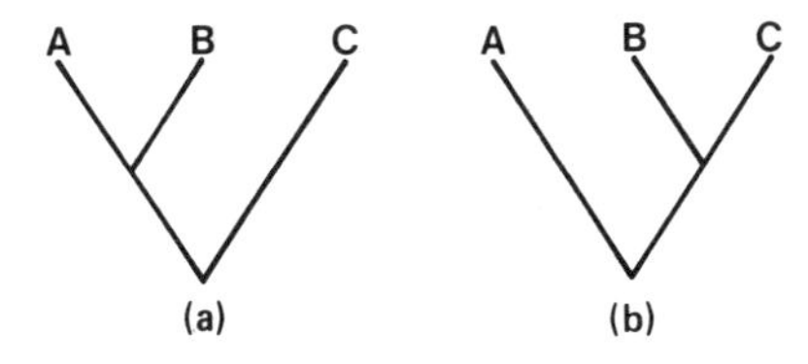

Fig. 3.1. Two hypothetical cladograms of three taxa or areas.

Not all cladistic analyses of taxa in a particular biota will yield area cladograms as straightforward as those in the above example; that is, nine taxa showing one set of relationships, and one taxon showing another. In the previous chapter, we considered some of the theoretical developments that led to the recognition of patterns in cladistic biogeography, and the methods that have been suggested to derive patterns from cladograms of taxa. In this chapter, we consider what constitutes a pattern, what constitutes a conflicting pattern, and what sorts of explanations one can propose for these patterns. We consider competing geological and ecological explanations, as well as differences in estimation of the relative ages of taxa. We also consider what predictions a pattern for a biota allows us to make about cladistic relationships of unstudied taxa.

Thus, the question we aim to answer in this chapter is: When looking at the real world, what does a cladistic biogeographer do?

We have chosen two related problems in biogeography to put the methodology into practice: relationships of the taxa in common among North, Middle, and South America, and the relationships of North American taxa that are part of the holarctic or boreal zone biota. In the first case, our only constraint in choosing groups to include in our analysis is that a cladistic hypothesis involving three or more taxa is available and that at least one member of the group occurs in each of the three areas, North, Middle, and South America which, together for convenience, we shall call the 'New World'. In the second example, we examine a set of cladograms available for North American taxa involved in holarctic relationships to see what information we can derive from them about the history of the temperate North American biota.

The present day distribution of plants and animals has been caused by a variety of phenomena: geological history, climate, extinction, individual dispersal, and introduction, to name but five obvious factors.

The last of these factors, introductions, is significantly different from the other four in our list because it represents a known or suspected recorded mechanism affecting a distribution pattern. Such distributions are of little interest to us because they require no hypotheses of explanation. For example, we need not consider here the *Eucalyptus* of California because they are known to be the result of introduction of a group endemic to Australasia. If we did not know *Eucalyptus* had been introduced into North America, its distribution might intrigue us.

The other factors in our list, for example climate (in its various manifestations, such as glaciation), obviously change distribution patterns over time. However, we do not know precisely what effect they have had, and how different their effects have been on different groups of plants and animals. For example, if a taxon did not exist in North America at the time of the Pleistocene glaciation, we would not expect its distribution pattern necessarily to resemble the patterns inferred from data on glacial pathways and the division of biotas. However, we may conclude that the pattern is at least as old as the Pleistocene, if this taxon, suspected not to exist at the time of glaciation because it has only post-Pleistocene fossils, did have a distribution pattern the same as other taxa in a general pattern of distribution concordant with glacial pathways. We may also suspect that we have underestimated the age of our first taxon and the fossils give only the minimum age of that taxon.

Thus, we emphasize the importance of the cladistic technique for pattern analysis, for only from the derived area cladograms can we know if a group conforms to a general explanation from distribution patterns of other groups. In effect, area cladograms are the most resolved summaries of relationships of areas for those groups for which we have reliable data. It is necessary, in searching for an explanation of the distributional history of a biota, to construct and compare cladograms of taxa to find the biogeographic patterns. Distributions that do not fit a general pattern have usually been explained by vicariance and cladistic biogeographers as random dispersals (e.g. Croizat *et al*. 1974; Rosen 1976, 1978; Nelson and Platnick 1981). Such incongruent patterns could equally have been caused by more complex geological histories or more recent ecological factors. Endler (1982) has challenged the whole of vicariance biogeography and cladistic analysis by claiming that concordance among cladograms indicates 'shared environmental effects' (p. 449) but not 'history'. As we use more and more disparate groups in cladistic biogeographic studies, such as plants as well as animals, the chance of our finding the same patterns caused by environmental factors diminishes. Later in the chapter we address some problems in interpreting patterns in terms of environmental or historical factors.

We maintain that all biogeographic studies must have at their base an historical biogeographic analysis. Initially, all taxa are treated equally and their cladograms are combined to derive a general distribution pattern. Assumptions about age of groups or their present or previous dispersal ability or activity have no place in deriving a general pattern, for it is precisely these assumptions we wish to test. We shall develop some of these arguments by examining the New World and then North America on its own.

3.2 SAME PATTERN: DIFFERENT TAXA

Cladistic analyses for most New World taxa do not exist—which is in fact the situation we face for any area of the world we may wish to look at in detail. The Caribbean region in particular, as well as Middle America in general, has figured prominently in biogeographic studies because of the complex geological history of the region (e.g. Rosen 1976) and because of the importance placed on the formation of a Panamanian land bridge as a dispersal route for mammals and other animals between North and South America during the Pliocene (e.g. Marshall 1979; Marshall and Hecht 1978).

3.2.1 Congruence

We examine available cladograms to see what patterns exist in the distributions among North and South America and their intervening connections. The degree of congruence among cladograms of taxa from North, Middle, and South America is the extent to which they agree.

Consider the three cladograms of Fig. 3.2. They are taken directly from published diagrams, or derived from statements made about group interrelationships. Figure 3.2(a) represents the relationship of three sympatric groups; a subgroup of New World characin fishes, heterandriine killifishes (see Figs 2.10, 2.11) and New World representatives of the plant genus *Magnolia*. The relationships of four taxa of each group are specified: a North American taxon is most closely related to a South American taxon. For the most part, these groups are found in subtemperate to tropical areas, and are absent from temperate southern South America.

Figure 3.2(b) is derived from a cladogram of genera of neotropical rivulid killifishes (Parenti 1981*a*, Figs 20 and 93). One primitive genus, doubtfully monophyletic, is found in North, Middle, and South America. A second is found in Middle America and northern South America. The rest of the rivulid genera are found in either northern or southern South America, and are related in a repeated dichotomous pattern. Again these killifishes are absent from temperate southern South America.

Figure 3.2(c) represents the area relationships of species within the possum-like marsupial genus *Didelphis* (see also Fig. 3.3). The North American species has representatives in Middle America; its sister group is a Middle and South American species. Together, these two species form the sister group of a third species, in temperate, southern South America (Gardiner, in Patterson 1981*a*).

What can we do to summarize the information in these cladograms so that we may make a general statement about the distributions of a group of characins, heterandriines, magnolias, rivulids, and some didelphids? We look first for the common elements in each of the three cladograms (Fig. 3.2(a)-(c)). North America is mentioned in all three, so that is one general area. Middle America is mentioned in all three but is divided into two spatially separate areas, which we will call Middle America 1 and Middle America 2. South America is mentioned in all three, yet it is divided into two areas, a northern and a southern, only in two cladograms (Fig. 3.2(b), (c)). Should we divide South America into two areas for our general pattern? Yes, if the object is to summarize everything we know about relationships of areas. No, if the object is to summarize what we know in common about the five plant, fish, and marsupial groups. Our purpose here is the latter, so our general pattern is that of Fig. 3.4, that is, South America is treated as one area. Our purpose could have been the former; we address such a problem in the next chapter. By treating South America as one area, we are not really losing information, just making our investigation that much more general.

What sorts of patterns would conflict with our general pattern of Fig. 3.4? Consider the area

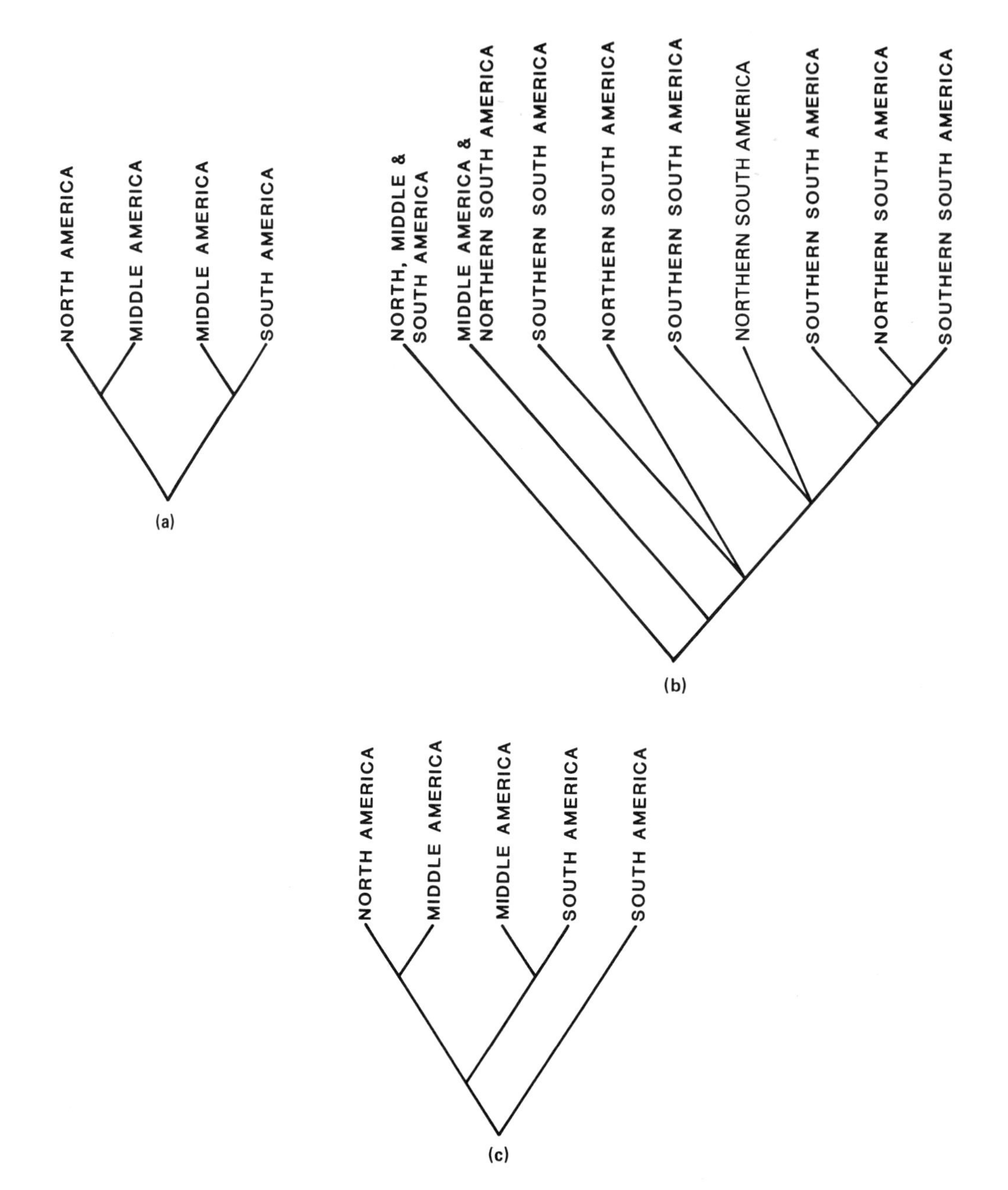

Fig. 3.2. (a) Cladogram of areas as specified by relationships within a subgroup of New World characiform fishes, heterandriine poeciliid fishes, and New World magnolias.
(b) Cladogram of areas as specified by relationships of genera of rivulid killifishes. (After Parenti 1981*a*.)
(c) Cladogram of areas as specified by relationships of species in the marsupial genus *Didelphis*. (After information from Gardiner in Patterson 1981*a*.)

Fig. 3.3. *Didelphus azarae* and young (Hudson 1892, p. 102.)

cladogram for fossil didelphids (Fig. 3.5(a)). North American and European taxa form a sister group, they in turn are most closely related to a South American taxon. Does this pattern conflict? No, because it contradicts none of the relationships of area in Fig. 3.4. What it can do is add another area, Europe, for possible consideration. Fig. 3.5(b) does not conflict with either Fig. 3.4 or Fig. 3.5(a).

An area cladogram such as Fig. 3.6 would conflict with our general pattern, but only if the taxa from South and North America were from the sub-temperate and tropical region, that is, if we were treating the same areas of endemism. The

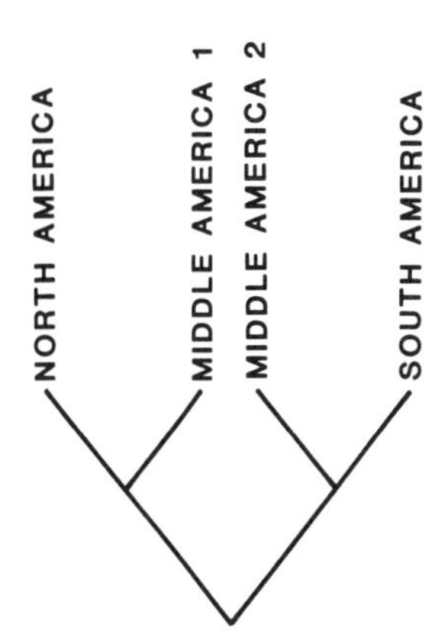

Fig. 3.4. General pattern of area relationships derived from cladograms of Fig. 3.2.

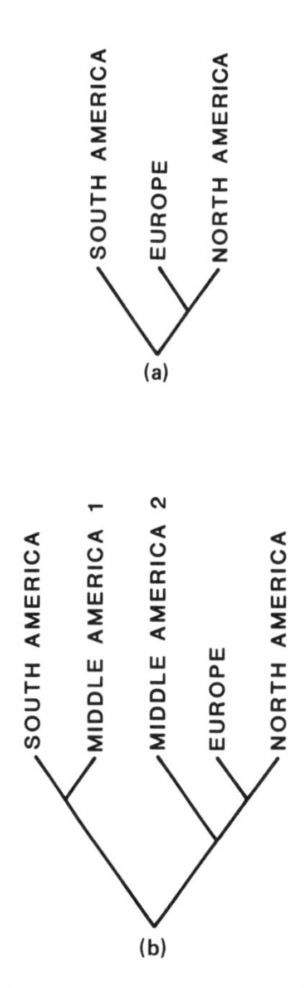

Fig. 3.5. (a) Cladogram of areas of fossil didelphids (after Patterson 1981*a*). (b) General pattern summarizing area relationships specified in cladograms of Figs 3.5(a) and 3.4.

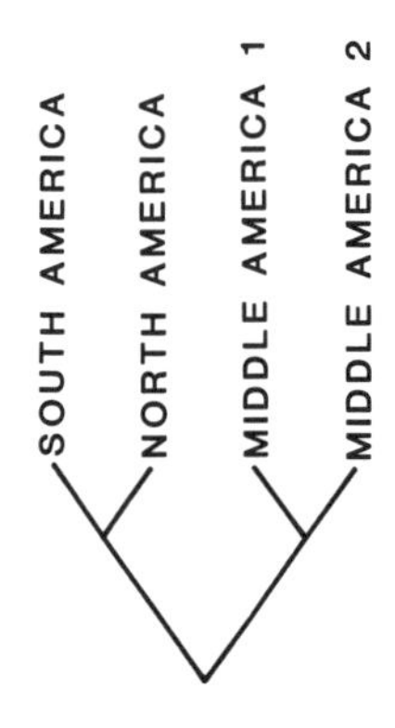

Fig. 3.6. One possible pattern that conflicts with the general pattern of Fig. 3.4.

pattern in Fig. 3.6 is supported for the cold-temperate regions of these continents (see Chapter 4).

3.2.2 General explanations

How does one explain a pattern such as Fig. 3.4? It has been maintained by many biogeographers that distribution patterns should be explained group by group; that is, no general explanation should be looked for because different groups have different dispersal abilities or are of different ages. The latter idea is often allied with strict interpretations of the fossil record and a reliance on the record to explain somehow historical distributions. It has been assumed that groups of plants and animals arose in different epochs and dispersed around the world as and when each could. For animal groups, the Paleozoic and Mesozoic periods are traditionally the Age of Fish, then of Reptiles. The Cenozoic is the Age of Mammals. This hypothesis of the spread of plants and animals over the world is also tied to ideas about a stable earth, with no continental drift as exemplified by George's statement; 'But it seems easier to make animals move round the world and fill permanent continents than to make the continents move round to collect them' (George 1962, p. 96).

The incorporation of almost universally accepted ideas of a changing geography has now altered the explanations given for particular plant and animal distributions. Speculation about the ages of groups relies on the assumption that a group is at least as old as its oldest fossil representative. Some biogeographers have emphasized that a fossil gives only the minimum age of a group (e.g. Croizat 1964; Patterson 1981*a*), but this idea has not yet been incorporated fully into biogeographic studies.

The distribution of the North, Middle, and South American mammal fauna has been studied extensively. South America has a distinctive mammal fauna, as notable for what it lacks (multituberculates, many eutherians) as for what it contains (marsupials, edentates, bats; McKenna 1980).

In the latest Mesozoic and the earliest Cenozoic, North and South America are believed to have been joined by a strip of land similar in shape and position to the present Middle American connection. Because fossil marsupials are known in North America from this time, it is assumed that they, along with representatives from other vertebrate groups, entered South America from the north across the land bridge. The land bridge was submerged soon afterwards, thus prohibiting immigration from the north until it again came into existence in the late Cenozoic. Hence, the South American mammal fauna is said to have evolved in 'splendid isolation' (Simpson 1980). South and North America became joined yet again by the land during the Pliocene, an event which is held to be responsible for the 'Great American Interchange' (see Webb 1976, 1978; Marshall 1979; Marshall and Hecht 1978; Marshall, Butler, Drake, Curtis, and Tedford 1979; Simpson 1980).

Briefly, the reason for belief in the northern origin of South American mammals is that, by and large, their fossils appear late in the record and are relatively advanced. Why are some mammalian groups not in South America? One frequent answer is '. . . They never reached South America because of some barrier, the most obvious kind being an ocean.' (McKenna 1980, p. 55.)

A corollary of this type of explanation is that any mammals arriving in South America between the late Mesozoic and the Pliocene are hypothesized to have dispersed and, most likely, from one of the closer Central American land masses (e.g. Marshall 1979).

These hypotheses have been presented to explain the distribution of New World mammals. Less attention has been given to the other groups supporting the pattern of Fig. 3.4. Characin fish are hypothesized to have dispersed into South America from the north (e.g. Novacek and Marshall 1976). Parenti (1981*a*) favoured the viewpoint for killifishes that their distribution pattern, at least in part, was best explained in terms of ancient geography. The distribution of marsupials has received considerable attention from biogeographers, although relationships have not been worked out satisfactorily (see Patterson 1981*a* for a review). Rosen (1976), in reviewing Caribbean biogeographic patterns in several groups by the vicariance method, emphasized that dispersal need not be an *a priori* assumption of the cause of a particular distribution, and that present distribution patterns need not necessarily have been caused by relatively recent (Quaternary) climatic events. Parts of distribution patterns undoubtedly have been caused by relatively recent and chance events, yet a common underlying pattern for several groups inhabiting the same areas requires a general explanation.

What then can we say about the pattern of Fig. 3.4? What sort of general explanation could fit different groups? What would such an explanation mean; what would one use it for?

A parsimonious explanation for the pattern (Fig. 3.4) follows. An ancestral biota is suggested by the pattern that includes North America, Middle America, and South America. But we must remember that the groups have primarily tropical associations, rather than temperate, so we may want to restrict the ancestral biota to exclude the temperate regions. There were three divisions of this ancestral biota, one to create a North America-Middle America 1 area and a Middle America 2-South America area, a second to create separate North America and Middle America 1 areas, and a third to create separate Middle America 2 and South America areas. These 'divisions' are vicariance events that have divided the ancestral biota. Whether these events represent actual movement of land masses or movement of the biota because of some shift in climate is unknown. Given the history of the region, it seems most likely that there was initially a continuous North and South American tropical biota that was disrupted as these continents separated, and the intervening Middle America was colonized from North and South America (see Rosen 1976). However, a concise picture of the geology of the Caribbean remains to be worked out.

What is important about this explanation is that

it is good for every taxon exhibiting the pattern. There is no need to formulate a separate explanation for each group, and no need to postulate migratory exchanges of mammals between North and South America. Individual explanations for the distribution of the groups would prevent the discovery of a single pattern.

3.2.3 Predictions

Having established a general pattern, we consider what predictions we can make to test in future studies. Predictions may concern at least four topics: cladograms of taxa, cladograms of areas, geological hypotheses supported by the patterns, and ages of biotas.

The first two, cladograms of taxa and areas, are intimately related. We predict that taxa found in the warm temperate and tropical regions of the New World are related in a manner that yields an area cladogram congruent with Fig. 3.4, that is, we predict that additional taxa would not conflict with the pattern. In this way we test one important aspect of our pattern: does the general explanation really help to describe the history of a region (here the New World), or is it unique to our five taxa, and therefore irrelevant to the other members of the New World biota? (We emphasize here that relationships of taxa in the endemic areas are predicted at any taxonomic level. We do not believe it is instructive to restrict ourselves to comparing relationships of, for example, genera or families of birds, fish, and mammals because many such categories are, at present, arbitrary).

Corroboration of the general pattern by more groups would support our interpretation of the history of the areas, that is North America shares a history with Middle America 1, South America shares a history with Middle America 2, and these two pairs together form an ancestral biota. We have greatly simplified the history of a geologically complex area, the Caribbean, in order to discuss our final prediction concerning ages of biotas.

Groups of plants and animals that share a cladistic pattern share a history. Implicit in this statement is the concept that all groups sharing the pattern are of approximately the same age, or at least all as old as the first vicariant event in the formation of the pattern—otherwise, they could not all have responded to the same events at the same time.

We see no special role for fossils in cladistic biogeography except to help in rejecting geological explanations for a particular pattern. Most taxa, such as rivulids, do not have fossil representatives. Of those that do, such as didelphids, some are relatively old, i.e. Pre-Palaeocene, and some are relatively young. Using fossil evidence alone, we can estimate the minimum age of didelphids as Pre-Palaeocene; we have no direct evidence for the age of rivulids other than Recent (see 3.3.2).

If we suppose the initial split causing the pattern of Fig. 3.4 was a Cretaceous rift between North America-Middle America 1 and South America–Middle America 2, we estimate the minimum age of the ancestral New World biota to be Cretaceous. Therefore, the estimated minimum age of rivulids and didelphids would be Cretaceous.

If one or several groups showing the pattern of Fig. 3.4 had Triassic fossil representatives, we could come to either one of two conclusions:

(a) the pattern is at least as old as the Triassic and we have underestimated the minimum ages of the other taxa; or

(b) the group with Triassic representatives remained unaffected by any postulated vicariance events until the Cretaceous, by which time other taxa had evolved, become sympatric with it, and all were affected together by the subsequent vicariance events.

3.3 DIFFERENT PATTERNS; DIFFERENT TAXA

In the foregoing example for the tropical New World, we have considered possible interpretations and predictions of a pattern exhibited by a number of unrelated taxa. In this section, we look at a group of Holarctic taxa for which there are conflicting patterns.

Before surveying patterns of Holarctic taxa, we consider one explanation for conflicting patterns, that our hypothesis for the phylogeny of a group is wrong. We have mentioned that biogeographic studies must be based on sound phylogenetic studies, but when faced with a taxon that does not fit a pattern, the temptation to doubt the phylogeny is strong. It is to a cladistic biogeographer's advantage to work with a cladist revision.

The derivation of the North American freshwater fish fauna has recently been given attention by cladistic biogeographers (Patterson 1981*b*).

Cladograms of areas for three fish groups—paddlefishes, bowfins, and umbrids (mudminnows) are given in Fig. 3.7. These are all modified from Patterson (1981*b*), and were derived from published information. The cladogram for umbrids includes two positions for the genus *Dallia* of Alaska and eastern Siberia, represented by two dotted lines, because the relationships of umbrids are debatable in this regard.

The different areas named in these cladograms are treated for convenience as four areas: western North America (including Alaska), eastern North America (including south-eastern North America and central North America), Europe, and Asia (including eastern Siberia).

When both eastern North America and Europe appear together on a cladogram, they may be sister groups (Fig. 3.7(b) for recent umbrids). A conflict arises when trying to decide which area, Asia or western North America, is most closely related to them. In bowfins (Fig. 3.7(c)) the answer is clearly western North America; however, for paddlefishes the answer is Asia; umbrids are unresolved.

We begin to see a problem in summarizing clearly the relationships among the four areas. In fact, the information from the cladograms of Fig. 3.7 is best summarized by the cladogram of Fig. 3.8(a); that is, there is an unresolved trichotomy among Asia, western North America and eastern North America-Europe.

However, there is possible further evidence on the relationships of these three areas that can help us resolve the problem. Lance Grande (personal communication) examined relationships of fishes of the Green River Eocene, and found that the general pattern is for western North American taxa to have a closer association (i.e. be more closely related) to Asian (or other trans-Pacific) taxa than with eastern North American or European taxa. Thus, a resolution of the area cladogram of Fig. 3.8(a) is presented in Fig. 3.8(b) in which Asia and western North America form sister areas, that together are most closely related to an eastern North American-European sister area pair.

We consider four possible explanations for this pattern:

1. The relationships proposed for some groups is wrong, and all North American taxa form a monophyletic group that is either most closely related to Asian or to European taxa.

2. One group of taxa was influenced primarily by vicariance events affecting that section of the North American biota, whereas the other group of North American taxa dispersed.

3. The two patterns reflect a real, separate history of western and eastern North America.

4. The two patterns reflect an older and a younger component of the biota, the older pattern caused by older vicariance events, the younger pattern by more recent events.

The first of these explanations, erroneous phylogenies, is of course always possible, yet very unlikely given that one set of taxa exhibit one pattern whereas a second set exhibits another. Therefore, we consider the two positions for North America in Fig. 3.8(b) to be real, and seek an explanation among the three other possibilities.

The second possibility, the idea that one pattern for North America is explained by dispersal and the other by a set of vicariance events that affected all the taxa in another part of the continent is contrary to what we predict about patterns; that is, that they have general explanations. Therefore, we are left with two possibilities for an explanation of the 'biphyletic' nature of the North American biota, which we discuss now separately.

3.3.1 Areas of hybrid origin

The decision to treat North America as two regions, an eastern and a western, reflects directly

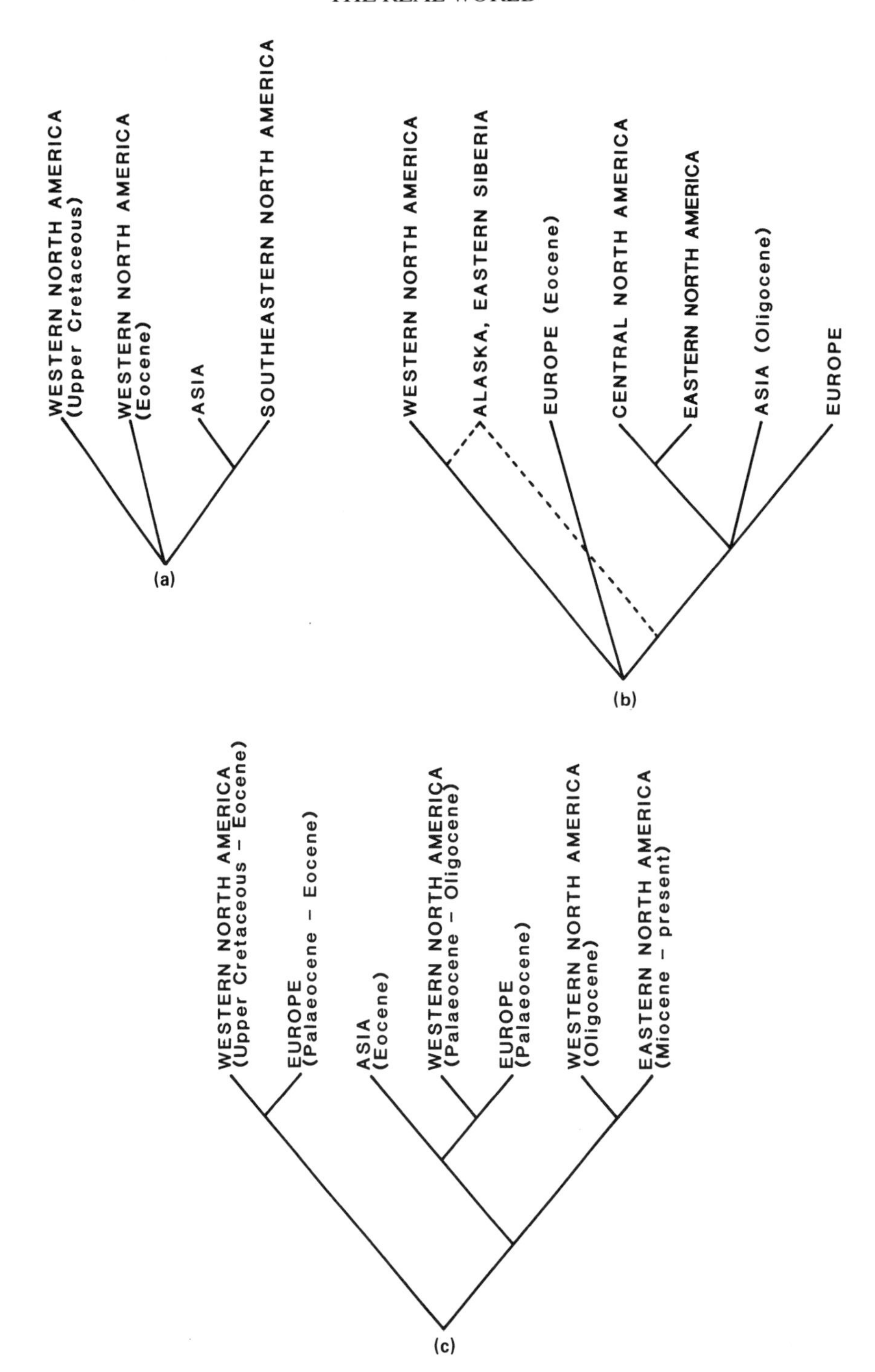

Fig. 3.7. Area cladograms along with ages of fossil representatives for (a) paddlefishes, (b) umbrids and (c) bowfins. (All modified from Patterson 1981*b*.)

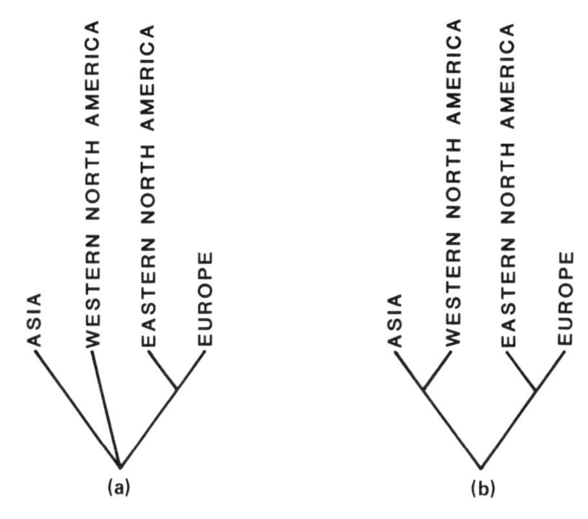

Fig. 3.8. (a) General pattern of area relationships derived from cladograms of Fig. 3.7. (b) General pattern of Fig. 3.8(a) with position of western North America resolved.

the findings on the relationships within the North American fish fauna. The division of the continent, however, is not clearcut because it may differ from group to group based on recent distributions.

For example, the umbrids of western North America are confined to Washington State, whereas the Green River Eocene fishes are from Wyoming, Utah, and Colorado. Thus, when we say that western North America is an area distinct from eastern North America we are referring to a very vague division of the continent. The pattern of Fig. 3.8(b) implies a separate history of the two areas, western North America sharing a basic component of its history with Asia rather than with eastern North America, which shares a basic component of its history with Europe. Since what we call North America is now, and for a long period has been, one continent, these 'basic components' of history, if they reflect independent geological histories of western and eastern North America, suggest the continent was separated into two or more areas a long time in the past.

Geological evidence that North America was at one time separated into two or more areas would support the idea that the pattern of Fig. 3.8(b) is directly the result of the geological history of the continent. The idea is not original to us; it has been hypothesized for some time that Asia and some part of western North America lay side by side. There are two possibilities: either the eastern and western halves of North America constitute at least two separate units fused together, or, at one time an epicontinental sea divided a single continent into two. There is geological evidence to suggest that the growth of North America has not been gradual but episodic. According to Jones *et al.* (1982) virtually the entire Pacific coast from Baja California in the south to the tip of Alaska in the north and extending up to 500 km inland is a series of prefabricated blocks grafted onto the main North America block. These blocks, known as 'accreted terranes', have been carried many thousands of kilometres east and north from their sites of origin in the Pacific.

Second, that an epicontinental sea divided the North American continent into western and eastern components from the upper Jurassic to the lower Cretaceous, especially during the upper Cretaceous and not before, or since, is geologically well established (see Howarth 1981 for review). We discuss this idea more fully in the next chapter, but emphasize here that the biological pattern suggests different histories for western and eastern North America since the time these fish have existed, and also perhaps predicts what geologists will find to be the case, rather than refutes or corroborates any one geological hypothesis that we would advance for the history of the continent.

3.3.2 Patterns of different ages

Suppose that there was no evidence for the independent geological histories of western and eastern North America, or that if there were it indicated a separation of the continent in the very distant past so that we thought it could have nothing to do with causing the existing biological pattern. We could consider that there is an older and a younger component to the biological pattern. As a hypothetical example, we could consider that one group of North American taxa, with closest relatives in Asia, had existed in western North America since the Cretaceous. Members of the taxa may have even been more widespread than they are now, and had their range reduced by subsequent extinction. A second group of taxa, relatively younger, in existence since the Oligocene, could be widespread in eastern North America, with closest relatives in Europe.

How viable is this explanation for our pattern? Relationships of western North American taxa with Asian taxa are at least as old as the Eocene (Fig. 3.7(c)). Relationships of eastern North American taxa with European taxa are at least as old as the Eocene as well, so it is impossible with Eocene data to resolve the general pattern (Fig. 3.8(b)) into two patterns, one older and one younger.

When we look further at the ages of the area associations of Fig. 3.7, we see that there is an old (Upper Cretaceous to Eocene) relationship between western North America and Europe (Fig. 3.7(c)). This does not conflict with the general pattern of Fig. 3.8(b), as we have discussed in deriving the general pattern. Nonetheless, it indicates some older association between western North America and Europe. This association could mean either that there was some ancient ancestral biota that included the two areas, or that in fact just part of western North America is associated with Europe and part with Asia. However, a decision to further subdivide North America in our biogeographic study can only be based on conflicting cladistic relationships of taxa endemic to those areas. Additional cladistic information for more North American taxa may require such a subdivision. However, we caution that geological evidence alone should not require subdivision of North America in a biogeographic analysis but only suggest an explanation for a biological pattern. Otherwise we would be letting geology rather than biology dictate our regions of endemism, and that would be unsound biogeography.

These predictions, that there are older and younger components to a biogeographic pattern, need to be corroborated by additional fossil evidence. A younger pattern automatically becomes the older pattern if a fossil of greater age than the older pattern is found. Available fossil evidence may vary from group to group and be a result more of vagaries of preservation and discovery than the actual age of the group. According to Patterson (1981*a*), with whom we concur, fossils have two uses in cladistic biogeography. First, they document extinctions, thereby by inclusion extend ranges of groups allowing new areas and new test groups to be brought in. Second, by giving minimum ages to groups, they ensure that comparisons between different groups are valid and permit a choice of geological events of different ages.

3.4 GEOLOGY AND THE CLADISTIC BIOGEOGRAPHER

Geological hypotheses for the history of areas of endemism have influenced biogeographers, more often than not, to such an extent that they interpret biological patterns solely in terms of prevailing geological theory or do not consider patterns within biotas, but interpret the distribution of a single taxon in terms of one particular geological explanation.

Geological hypotheses for the histories of areas of endemism are no more or less close to the truth than the history of a biota inhabiting those areas as presented by a biogeographical pattern. As Croizat (e.g. 1964) has repeatedly said, the biological patterns exist whether or not they agree with the geology.

We assume that, on some level, information on the history of areas derived from biological and geological data should agree. However, that the biological relationships and their associated area relationships should stand as evidence independent of geological relationships of the areas is one of the tenets of cladistic biogeography. Nonetheless, in searching for explanations of distribution patterns, biogeographers are continually intrigued and influenced by geological evidence. We consider here some of the ways in which geological evidence has been used in biogeographic studies, and ways in which it may help explain a pattern.

3.4.1 Cladograms of taxa

What refutes a cladogram of taxa; in other words, what evidence would make us change our hypothesis of a general phylogeny of a group? The addition of new characters, or doubt about some characters already used to construct the cladogram, but not details of the geography. If a cladogram of taxa indicates that the taxon in area A is more closely related to the taxon in area B than either one is to the taxon in area C (as in, for example, Fig. 3.1(a)) and the prevailing geological hypothesis indicates that area B shares a history with area C but not with A (as in, for example, Fig. 3.1(b)), we would not conclude that the geology refutes our biological analysis. Furthermore, and perhaps more important, we must always try to test a pattern in biogeography, not test individual cladograms. In order to begin to construct general explanations for distributions within biotas we must examine at least two and preferably several groups of taxa. Examining the cladogram of relationships within one taxon and comparing it to the prevailing geological hypothesis is analogous to trying to decide on the basis of an organism's vagility whether or not its distribution is the direct result of random dispersal. It is a fruitless 'test' of a pattern where no pattern may exist.

For example, consider a case in which the geology of a region is fairly well-known, or even in which recent geological events are known with certainty because they have occurred within recorded history, such as the disruptions caused by the recent earthquake into areas A1 and A2, we might expect the taxa in areas A1 and A2 to be sister taxa. We might carry out a cladistic analysis of four taxa, one each from areas A1 and A2 and two other areas B and C. If we found that cladistically, taxa in A1 and B were sister taxa, and A2 and C were sister taxa, and both pairs together formed a monophyletic group, there might be several explanations for this pattern of relationships:

(a) our cladogram is wrong, perhaps based on characters that are environmentally determined and do not accurately reflect the phylogeny; or

(b) our cladogram is correct, but has nothing to do with the recently recorded geological events.

How could we test these two alternatives? The first alternative seems attractive since we might feel intuitively that if an area has been divided, the taxa in its subdivisions should be most closely related. If they are not, we might consider reasons for our erroneous cladogram. However, we stated previously one condition under which a cladogram might be doubted: if the taxa exhibiting it do not conform to a particular pattern. In this hypothetical case, however, we as yet have no pattern because we have looked at the relationships within only one group. Thus, the particular characters used to construct the phylogeny may be in doubt for other reasons, but not because they gave us a cladogram that does not correspond with one set of geological events. (See Endler 1982 for an alternative discussion).

To test the second possibility, we may search for some other, older set of geological events that correspond with the cladogram, and claim that these earlier events caused the inferred phylogeny, but we may never find such an alternative geological hypothesis. More important, proposing or not proposing such a hypothesis would not constitute a test of the phylogeny.

We have only one way of evaluating the second alternative: to search for a pattern among the members of the biota of areas A1, A2, B and C. If we constructed cladograms for five different groups in the biota and they were all the same as for our first group, that is taxa in areas A1 and B were in sister taxa, as were taxa of areas A2 and C, we would have discovered a general pattern for the biota. What caused this pattern? We may search for an explanation and find one that is suitable, such as, for example, our initial area A is like North America, one that should be treated as two (or more) areas of endemism for biogeographic studies. Whatever explanation we can find, we know that our initial prejudice about the affect of a known geological event (earthquake) could have led us to believe something that had little support. It is unjustifiable to say that a major prediction of cladistic or vicariance biogeography is that '... concordant cladograms should result from concordant vicariance sequences ...' (Endler 1982, p. 450), because histories of areas are complex and involve sequences which we may never recover in any geological study. Furthermore, every vicariance event will not result in a dichotomy in a cladogram. Any biogeographic study that ignores the analysis of pattern is faulted because it precludes the search for a general explanation.

3.4.2 Cladograms of areas

What refutes a cladogram of areas? Up to this point, we have considered cladograms of areas that were derived directly from cladograms of taxa. A cladogram of areas for the hypothetical biota of the preceding section would require that areas A1 and B be sister areas, and A2 and C be sister areas. Another way of deriving a hypothesis for the history of the areas would be to ask a geologist to construct one. Because the configuration of areas (including our ideas on what constitutes a continent) changes over time, the geologist may produce several hypotheses, each one for a different period, but none in conflict with the other; that is, no one hypothesis precluding the existence of another. We may decide that a geological event separating areas A1 and A2 did not produce the general pattern, and that we were perhaps wrong in thinking of area A, which may have been a continent such as North America before the geological event as one homogeneous area. The geologist may look further into the history of these areas and discover that areas B and C are more alike than A1 or A2 in a certain set of characteristics, and suggest that because of these similarities, B and C share a history that A1 and A2 do not. However, consider the possibility that because areas A1 and A2 have been joined as a continental land mass, they now have similar geological characteristics. What the geologist has discovered B and C share may be those characters that A1 and A2 no longer express because the characters have been modified. What B and C share are possibly older geological characters that would not form the data base for the construction of a geological cladogram.

Do geologists produce geological cladograms based on derived geological data, or phenograms based on overall similarity of regions? Biologists have little or no expertise by which to judge the geological data in any one case. Surely in some cases, the basis on which two areas are hypothesized to share a close history are derived characters analogous to such characters in biological studies. How significantly changes in the suspected primitive or derived nature of geological data affect current geological hypotheses will only be known in time. Any geological hypothesis is to be approached with the same caution or scepticism one might have for a phylogenetic hypothesis; they should both be subjected to rigorous test.

But can a geological cladogram be tested with a biological cladogram? No, say the majority of vicariance and cladistic biogeographers (e.g. Rosen 1978; Nelson and Platnick 1981; Patterson 1981*a*). Biological and geological hypotheses have been termed 'reciprocal illuminators' (Rosen 1978); they enhance but do not test one another.

If we look again at the example of the taxa in Areas A1, A2, B, and C, we understand that a cladogram can only be tested with additional biological data or a reassessment of the data used to produce that hypothesis. Similar tests can be made of geological cladograms by reassessing geological data or using new data. The failure of a biological cladogram to agree with a geological cladogram for the same areas of endemism should in no way diminish our belief in either, but it should make us scrutinize the two sets of data critically.

If we believe that the world and its biota evolved together, then we would expect our geological data to correspond with our biological data at some level. In so far as cladistic analyses of plants and animals have only been applied rigorously to a small percentage of groups and to an unknown extent in geology, our expectation may not be realized for generations.

3.5 CONCLUSIONS

In surveying some of the practical problems one might face in carrying out a cladistic biogeographic analysis we have touched upon those topics which have traditionally plagued biogeographers (as, for example, estimating the age of a group) and presented ways in which a cladistic biogeographer might approach the solution of these problems.

Cladistic biogeography centres on historical explanation for distributions. Because no one biogeographic study touches on all of the problems, we have discussed both real and hypothetical examples. Conclusions from such a variety of examples are diffuse, so we present a summary of the major principles of this chapter as a list.

1. Any biogeographic analysis (whether primarily historical or ecological) must consider the analysis of pattern.
2. Patterns, congruent cladograms for various unrelated groups, are what need to be explained in biogeographic analysis, not the distribution of individual groups.
3. Historical biogeographic analysis constitutes the search for pattern, whereas ecological analysis constitutes the search for individual responses to relatively recent events (viz. nonconformity to a pattern).
4. A vicariance event is anything that disrupts a biota (e.g. divides it into two or more subunits).
5. Vicariance events may be geological (e.g. an uplifted mountain range), climatological (e.g. a glacier), or any other factor that disrupts a biota.
6. Biogeographic analyses must be based on sound phylogenetic studies.
7. Conformation to pattern is of primary importance in biogeographic analysis. The dispersal capabilities of individual taxa are secondary, and become irrelevant if all the taxa conform to a particular pattern.
8. Nonconformity to a pattern may be the result of random dispersal, ecological variation or conformity to a different pattern.
9. As an explanation for a distribution pattern random dispersal is unique to a taxon, not to a biota.
10. The present distribution of plants and animals in the world has been caused by a variety of phenomena (including geological history, climate, extinction, dispersals, and introductions).
11. A biogeographic pattern is a summary of the area relationships specified by congruent cladograms of taxa.
12. All taxa, regardless of their presumed age or inferred dispersal ability, are treated equally in deriving biogeographic patterns.
13. The amount of congruence among cladograms of taxa is the extent to which they agree, that is, the extent to which they support a particular pattern.
14. An area initially treated as one area in a biogeographic study may be subdivided into two or more areas if the relationships within the biota dictate such a division (e.g. if all the taxa of a particular group in the initial area do not constitute a monophyletic lineage).
15. A pattern for a biota conflicts with another pattern for the same biota if one or more of the area relationships in the first pattern is contradicted in the second.
16. A fossil representative gives a minimum estimate of the age of a group.
17. One estimate of the age of a group is the minimum age estimated for the oldest taxon in the pattern which it supports.
18. A general explanation for a pattern applies equally to all taxa supporting the pattern.
19. Finding a pattern for a biota predicts that other taxa in the biota will also conform to the pattern.
20. Taxonomic level (rank) of a particular group in the biota is irrelevant.
21. Groups of plants and animals in the same area (a biota) that share a cladistic pattern share a history.
22. Fossils have no special role in cladistic biogeographic studies expect to help in rejecting

geological hypotheses that we suspect may have caused a pattern.

23. Geological hypotheses of area relationships are no more or less reliable than area cladograms derived from biological data.

24. Geological hypotheses do not test biogeographical patterns.

25. On some level, the history of areas derived from biological and geological data should agree; however, area relationships based on biological data can be independent of geological relationships of the same areas.

26. A cladogram of taxa can be refuted by other characters, especially if constructed originally on wrong or doubtful characters, but not by geology.

4 A NEW VIEW OF THE WORLD

4.1 INTRODUCTION

The integration of cladistics with biogeography has developed out of the recognition that theories of biogeography, as well as those of systematics, must be testable to be scientific and to be incorporated into theories of earth history. But we gain little from this integration of the two fields unless it can tell us something new about the world.

For over a century naturalists as well as systematists have recognized, proposed, and debated numerous theories on the origin of the similar biotas of the highlands of tropical New Guinea, subtropical New Caledonia, and the temperate areas of southern South America, New Zealand, Tasmania, and Australia, sometimes also including the southern tip of Africa (see, for example, Darwin 1859; von Ihering 1900; Eigenmann 1909; Hooker 1860 and summarized in Turrill 1953; Hubbs 1952; Darlington 1957, 1965; Croizat 1952, 1958, 1964; Craw 1982). Along with the fossil biota of Antarctica, these areas comprise what biogeographers have termed the austral zone. Because the derivation and distribution of the austral zone biota continues to be the focus of biogeographic studies (e.g. Darlington 1965; Keast, Erk, and Glass 1972; Good 1974; Diamond 1982; Craw 1982) we have chosen the austral zone as an example of one large, well-known region to which methods of cladistic biogeography can be applied.

Examples of groups with austral distributions, in whole or in part, include southern beeches of the genus *Nothofagus* (Figs 4.1 and 4.2), galaxiid (meaning star bangles, like the Milky Way) fishes (McDowall 1964; Rosen 1974), mordaciid (biting) lampreys (see map in Berra 1981), pines of the genus *Araucaria* (e.g. Monkey puzzle and hoop pine) and many other plant groups as listed in Du Rietz (1940) and Thorne (1972).

At the same time, biogeographers have been debating theories on the origins of amphitropical distributions; that is, distributions characterized by a close relationship between taxa in the north temperate zone and taxa in the austral zone (Darlington 1965). Such distributions are often also termed bipolar, amphitropical, or antitropical.

Examples of groups with amphitropical distributions include many remarkable Coleoptera groups (such as the devil's coach horse beetles, Gymnosine staphylinids; Hammond 1975), Carabid beetle groups (e.g. Broscini, Derodontidae, Byrrhinae, Nemonychidae; Crowson 1980), the spiny-finned fish family Percichthyidae (Fig. 4.3), crowberries of the genus *Empetrum*, lampreys, midges (Brundin 1966), a subgroup of the silverside fishes, beeches of the family Fagaceae (Humphries 1981), and the eyebright genus *Euphrasia* and other plant taxa as indicated in Du Reitz (1940) and Thorn (1972).

Dispersalist (Darlington 1965; McDowall 1964), vicariance (Rosen 1974), panbiogeographic (Croizat 1964), and cladistic (Humphries 1981) biogeographic studies have been carried out, proposing theories to explain the distribution of one or more animal or plant taxa inhabiting land masses of the austral zone. Whereas these discussions disagree on methods of analysing austral distributions they have one common element: limited consideration of how the origin of austral zone distribution is related to the origin of amphitropical distributions.

It is our purpose in this chapter to integrate fully these two closely related problems in biogeography to create a comprehensive hypothesis for the origin of global biogeographic patterns, and in doing so exemplify the methods of cladistic biogeography.

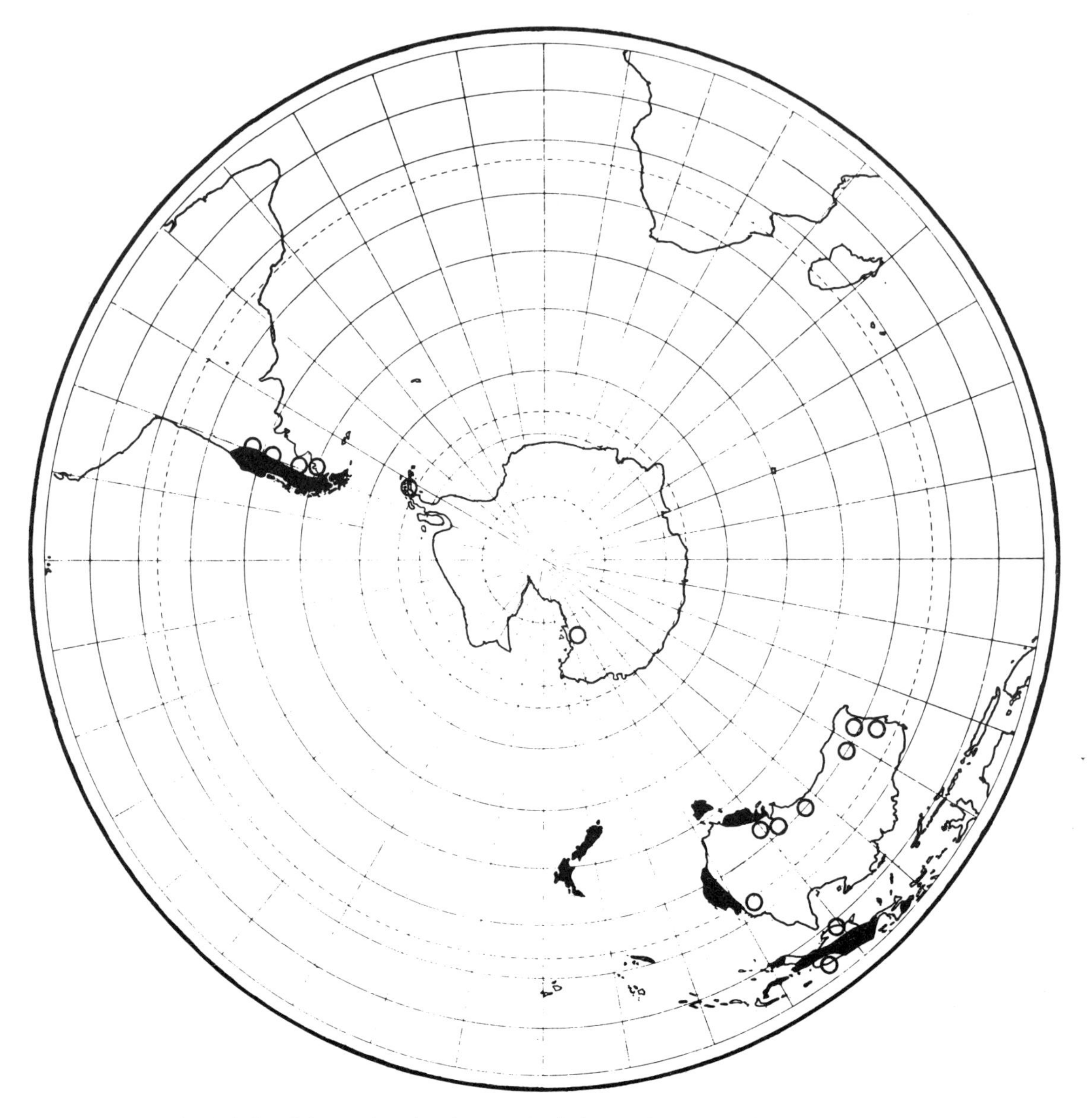

Fig. 4.1. Distributional limits of the southern beech genus *Nothofagus* (after Humphries 1981). Areas inhabited by recent species are blackened; dots in circles represent localities of fossils.

4.2 TROPICAL VERSUS AMPHITROPICAL

An amphitropical distribution, by definition, is one in which taxa are present in the northern and southern hemispheres, in particular the boreal and austral zones, but absent from the tropics (e.g. Darlington 1965; Hubbs 1952).

Conversely, pantropical distributions are those characterized by taxa in tropical regions of both hemispheres, but with no representatives in the cold-temperate austral and boreal zones. Pan-

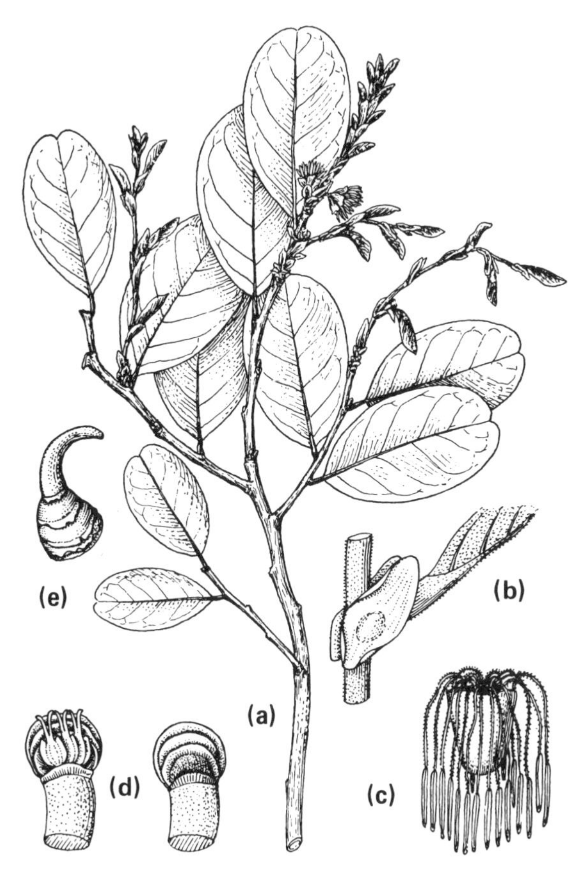

Fig. 4.2. *Nothofagus recurva* van Steenis from Papua New Guinea (van Steenis 1953, Fig. 5).

tropical distributions are approximated in numerous animal groups (for example killifishes, cichlids, and nandids among fishes, plethodontid salamanders, and sungrebes) as well as plant groups (for example, the nutmegs, Myristicaceae, Indian bean tree family, Bignoniaceae, and ebonies (*Diospyros*); see Fig. 4.4).

The boundary between the tropical and boreal zone in the northern hemisphere and between the tropical and austral zone in the southern hemisphere is not precise. Some groups characterized as fundamentally tropical, such as *Diospyros* (Fig. 4.4), coexist with typically boreal groups, such as the bog species, *Scheuchzeria palustris* (Thorne 1972) in north-eastern North America. Nonetheless, a taxon can generally be classified as a member of a tropical north and south temperate or bipolar group because of its relationship to other taxa and the distribution of those taxa.

A question that biogeographers have asked since the distinction between tropical and north and south temperate groups was noticed is: how did the disjunct north and south distributions come about? Answers to this question have involved climatic change (Darlington 1965; Hubbs 1952; Rosen 1974) long-distance dispersal from northern North America (Raven and Axelrod 1972; Thorne 1972) and bipolar differentiation (Du Rietz 1940).

Darlington summarized his thinking on the origin of amphitropicality as follows:

> All together, a large proportion of the plants and animals of the southern cold-temperate (austral) zone are involved in amphitropical distributions at one taxonomic level or another, and they include many of the most characteristic southern groups that are supposed by some persons to have dispersed by means of continental drift or antarctic land bridges. But the amphitropical pattern cannot be primarily a product of ancient geography. No biogeographers would seriously suggest (or would they?) that existing north and south temperate areas once formed a single land mass entirely separated from existing tropical areas, and that the amphitropical groups of plants and animals are still distributed according to the ancient division of land. Climate apparently must be primarily concerned in formation of the amphitropical pattern. (Darlington 1965, p. 130)

The importance of climate in determining where an organism can live today is self-evident; a warmth-adapted organism cannot disperse into a cold region and survive. If it does disperse and survive, evolutionary theorists would say it is both warm and cold adapted, and that climate has little to do with determining the formation of zones of distribution on a global scale. However, the importance of climate in determining how distributions came about is not self-evident, particularly when the relationships of taxa within amphitropical groups, and their relationship to tropical groups, are studied.

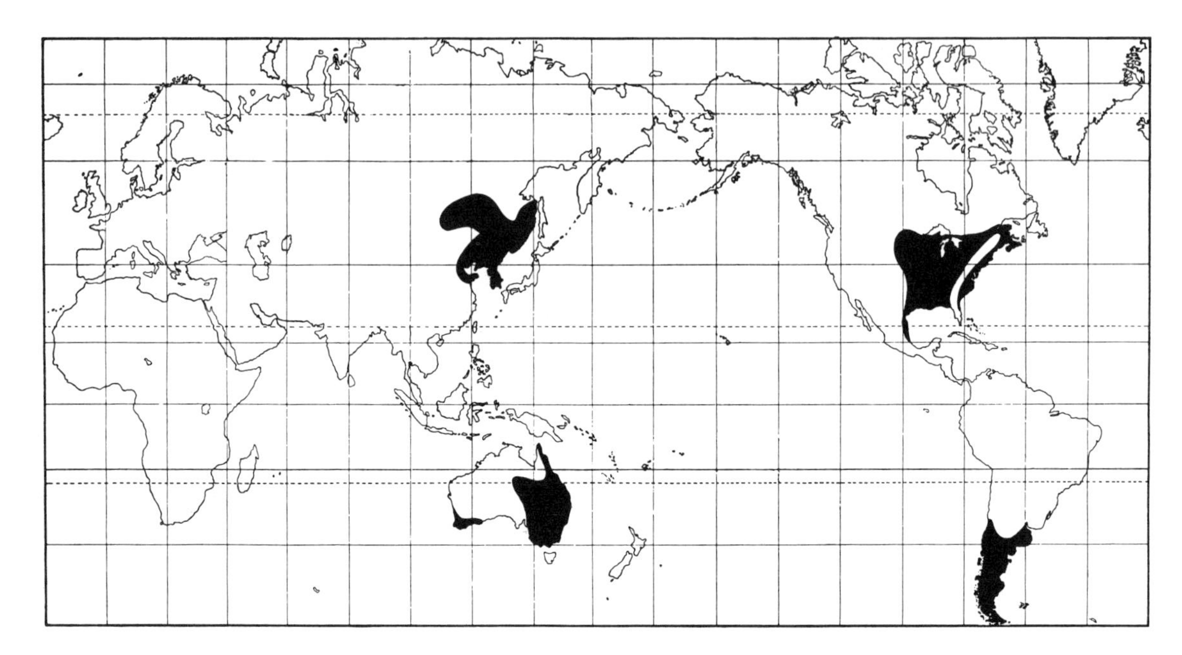

Fig. 4.3. Distributional limits of the percomorph (spiny-finned) fish family Percichthyidae. (After Berra 1981.)

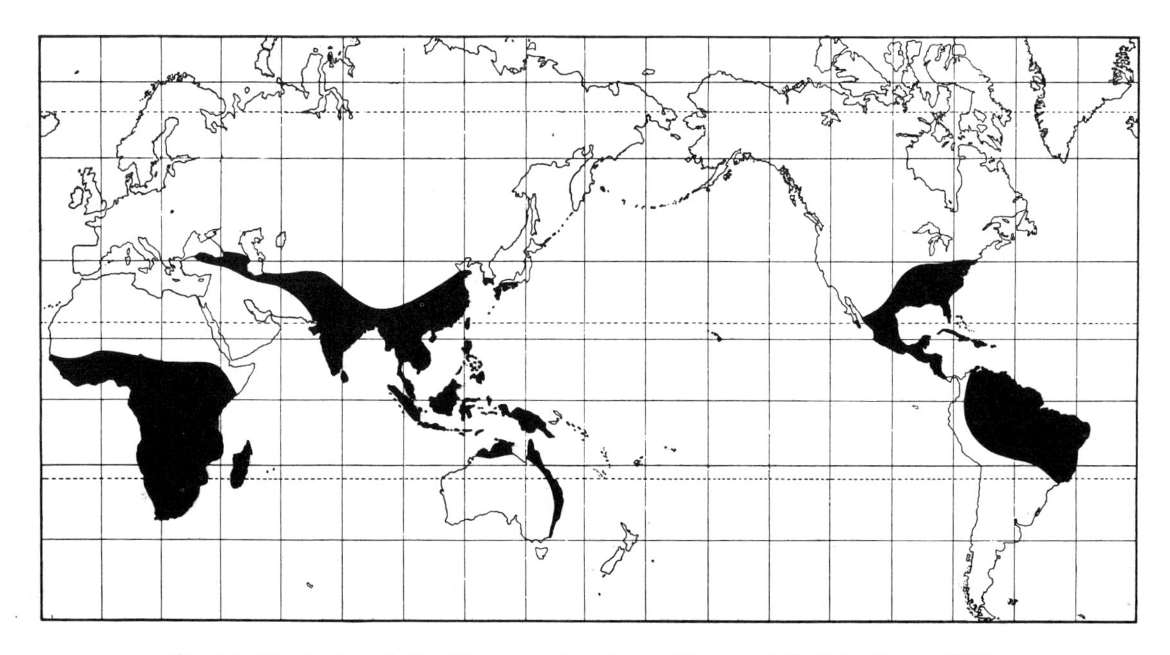

Fig. 4.4. Distributional limits of the pantropical ebonies *Diospyros* (s.l.). (After Thorne 1972).

The idea of bipolar differentiation, that is the independent formation of similar groups of taxa at either pole, was postulated initially by numerous biologists in the first part of the nineteenth century (see Du Rietz 1940 for a comprehensive review). As the theory of evolution became more widely accepted, it was proposed instead that these biotas were in fact closely related. Hence, long distance migration from one pole to the other was suggested and still remains a popular hypothesis explaining amphitropicality (e.g. Thorne 1972).

Independent bipolar derivation and migration from one pole to the other by members of each bipolar group were considered to be highly unlikely, and parts of the two theories were combined into one comprehensive theory of cosmopolitanism by Du Rietz (1940). Rather than postulating individual migrations, he stated:

> ... it seems to be equally possible for a genus or any taxonomic unit to differentiate or 'crystallize' out of its more polymorphic ancestral syngameon simultaneously over a very large area. If such a very polymorphic syngameon was once distributed both in the North and the South as well as over connecting transtropical highland bridges, it would be unnecessary to assume any transtropical migration for bipolar units later differentiated and isolated within this syngameon. And even the very polymorphic syngameon out of which they were differentiated may have flowed as a broad stream from previous syngameons with a similar distribution, so very old that any speculations as to whether they first came from the North or from the South would be futile. (Du Reitz 1940, pp. 230–1.)

This concept of ancestral cosmopolitanism and separation from the tropical zone by transtropical highland bridges is similar to Darlington's theory of zonation by climate because it implies the near exclusion of bipolar groups from the tropics. However, neither Darlington nor Du Reitz looked at relationships of tropical zone taxa to the bipolar taxa, therefore they did not address the question of how monophyletic groups of tropical taxa came to be wedged in between the sister taxa of the austral and boreal zones.

Amphitropical disjuncts are numerous in the literature. Of living floras, for example, Thorne (1972) indicates that at least 65 plant genera occur north and south of the American tropics. Amongst these are *Agoseris*, *Amsinckia*, *Bahia*, *Clarkia*, *Gilia*, *Haplopappus*, *Plectritis*, and *Schedonnardus*. Thorne also mentions that a dozen or more plant genera show bipolar disjunctions and amongst these are *Armeria*, *Empetrum*, *Hippuris*, *Koenigia*, *Phippsia*, and *Primula*.

Cladograms of relationship have not been worked out for many groups with amphitropical distributions, and few or no such hypotheses were available to Darlington when he produced his monograph on the distributional history of the southern end of the world, or to Du Reitz when he speculated on the origins of bipolar distributions. However, in the past two decades, cladistic hypotheses have been proposed for several amphitropical plant and animal groups. We examine some of these in order to arrive at an explanation for amphitropicality.

One group that has been the focus of numerous biogeographic studies is the southern beeches, the genus *Nothofagus* (Fig. 4.2; see Humphries 1981 for a comprehensive review). They comprise part of the widely distributed family Fagaceae. *Nothofagus* is confined to the southern hemisphere in what we have already referred to as a typical austral distribution (Fig. 4.1). Its sister genus *Fagus* occurs in the north temperate zone in North America and Eurasia. Humphries (1981) produced a cladogram of the areas inhabited by *Nothofagus* in the austral zone and *Fagus* in the northern zone (Fig. 4.5). *Nothofagus* species of the Patagonian Andes (South America) are most closely related to a group including representatives in New Zealand, Tasmania, and Australia, which in turn are closely related to a second group of Australian as well as New Guinean and New Caledonian taxa.

The cladogram of relationships within *Nothofagus* and *Fagus* has properties that are found in groups with amphitropical distributions. One is that there is a monophyletic group of austral taxa whose sister group is the monophyletic north temperate taxon. This is seen again, for example, in

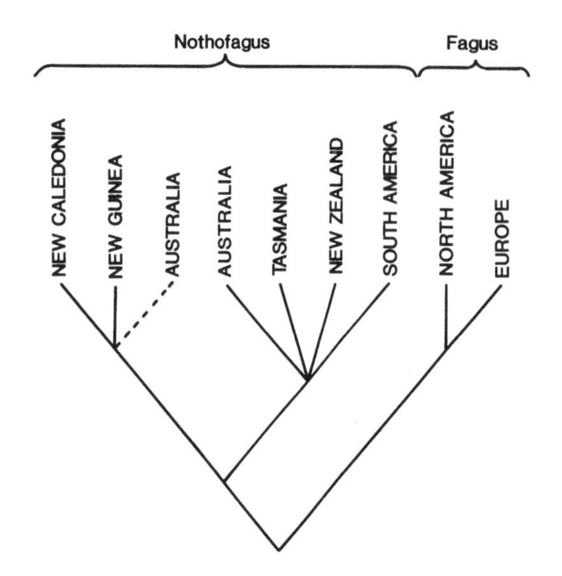

Fig. 4.5. Area cladogram of *Nothofagus* of the austral zone and *Fagus* of the boreal zone. (After Humphries 1981.)

midges of the family Chironomidae (see Fig. 4.6). Brundin (1966), in his monograph on the relationships and distribution of certain chironomids, established their status as a classic example of an amphitropically distributed group. Chironomids are represented in South America by two sympatric, not closely related subfamilies, the Diamesinae and Podonominae. The relationships among the areas inhabited by both subfamilies can be summarized in a single cladogram (Fig. 4.7).

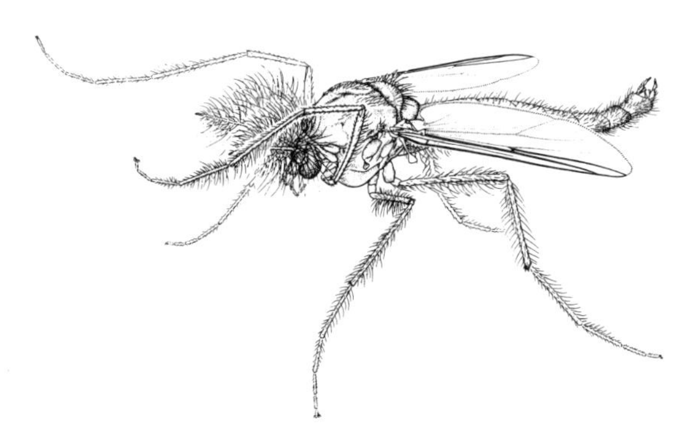

Fig. 4.6. Male of *Chironomus plumosus* L. (Oliver, in McAlpine, Peterson, Shewell, Teskey, Vockeroth, and Wood 1981, p. 423).

Midges of the Central Andes have their closest relatives in the Patagonian Andes. This group in turn forms a trichotomy with representatives in Australia and New Zealand. South Africa forms the sister area of these three. The austral midge taxa together comprise a monophyletic group. They are the sister group of the monophyletic boreal taxa of Asia, North America, and Europe.

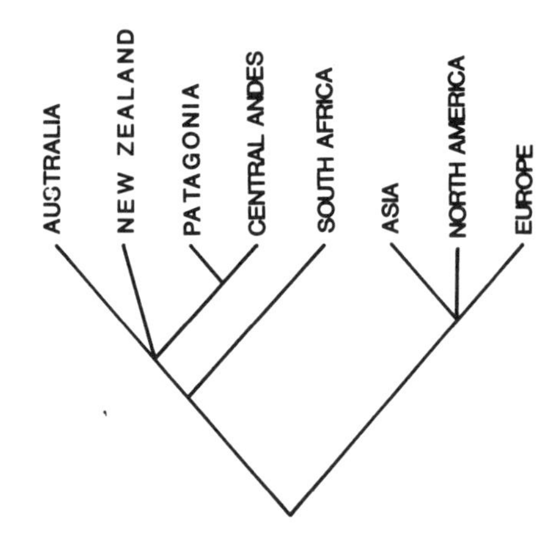

Fig. 4.7. Area cladogram of both podonomine and diamesine midges. (After Brundin 1966.)

(Midges of the Central Andes might be considered tropical rather than austral; however the Central Andes will be considered here along with austral zone distributions because of the close relationship between Central Andean and Patagonian taxa.)

The general congruence of the north-south relationships and distribution of two groups of midges and the beech group invites an explanation for this distribution pattern. At the same time, an explanation is invited for the presence of midges in South Africa with an austral relationship and perhaps also for the absence of beeches in that area.

First, we review three current theories of earth history in order to see if past land connections can in any way explain the pattern.

4.3 PANGAEA, PACIFICA OR AN EXPANDING EARTH?

The existence of an ancient supercontinent Pangaea, comprising all known global land masses that split and drifted to form the current configuration of the continents, is a theory that was proposed formally by the German scientist Alfred Wegener (1915). Subdivision of Pangaea, starting in the Mesozoic and continuing to the present day, has been the basis of the theory of continental drift, modified by geologists into a comprehensive theory of crustal evolution known as plate tectonics. Reluctance among geologists and biologists until recent decades to accept the concept of an unstable crust is now a classic story of resistance to new interpretations of earth history (see Nelson 1978*b*; Hallam 1973).

Darlington (1965), in providing the most recent pre-plate tectonics review did not accept the idea of a former supercontinent, and considered only that South America and Africa were joined at one time in the past; this he believed primarily because of matching between the present Atlantic margins of the two continents. In nearly twenty years since Darlington's monograph on the southern end of the world, the idea that continents change their relative positions, explained by theories of plate tectonics, has become widely accepted. Today, there are few geologists or biogeographers who doubt its reality, although the field is still alive with debate about the sequence and timing of continental separations and collisions.

There are two more speculative alternatives to Pangaea, which is itself a model based on the idea that the earth has essentially had constant dimensions. The second theory includes explanations for the apparent distributions of continental fragments around the Pacific margins. The third theory is one which argues for an 'expanding' earth to make better fits of the continents during the Triassic and Jurassic to overcome the problem that too much of the earth's surface has been swept under the continental carpets by subduction than present-day facts allow. The second theory (Jones, Cox, Coney, and Beck 1982) suggests that the major continents do not grow slowly and steadily, gradually accumulating 'rings' of rocks around the margins. Instead, the growth of continents is episodic. Jones *et al.* (1982) describe geological and geophysical evidence to suggest that, on the western coast of North America for example, huge pre-fabricated blocks are grafted piecemeal onto existing continent margins. Almost all of the Pacific coast of North America, from California to Alaska, consists of twenty or more added continental blocks (terranes) which were carried thousands of kilometres east and north, originating from a remote Pacific basin site. The dimensions of these terranes ranged from hundreds to thousands of kilometres long.

Jones *et al.* (1982) give no precise clues as to the origin of the terranes but there seem to be two explanations for their present-day existence. Tozer (1982) for example, has a palaeogeographic theory based on the distribution of Triassic marine fauna. During the Triassic, the western USA was interpreted as a tectonically quiet coast bordered by an open ocean. Well offshore there were a series of volcanic archipelagos shedding sediments into adjacent basins. The islands were within 30 degrees of the Triassic equator and extended offshore for about 5000 km, to the spreading ridge directly ancestral to the East Pacific rise. The geography west of the ridge was considered to be similar. During the Jurassic, new crust generation at the ridge pushed some of the islands into the North America plate, some to South America, others westwards to Asia. Tozer suggests also that New Zealand, northern New Guinea and New Caledonia were at a latitude of 30° north in the Triassic. The terranes of the western cordilleras of North America, he assumed, reached the North American plate before the end of the Jurassic.

This second explanation is a general theory comprising the former land mass and break-up of a southern supercontinent, Pacifica, now comprising bits of terrane distributed around the Pacific ocean margin. As its name implies, Pacifica was a hypothetical continental mass situated in the South Pacific, the initial rifting of which opened the Pacific Ocean. Supporters of this theory propose it in part as an alternative to the Pangaean model of continental break-up (Nur and Ben Avraham 1981). Thus, under the theory of Pangaean break-up, the Atlantic was opened by rifting continents; whereas under the theory of Pacifica break-up the continents were joined in some other fashion, with the Pacific formed by rifting continents.

In the third major theory Carey (1976), Owen (1976), and Shields (1979) argue that most Pangaea reconstructions assume an earth of modern dimensions. For them such reconstructions produce major continental fit anomalies in the Arctic, Caribbean, Mediterranean, and South-east Asia. They suggest that ocean floor spreading history of these areas and adjacent oceans indicate that they have grown by areal expansion since their initiation. Reconstructions for the Cenozoic eras for an earth of constant dimensions assume subductions to account for the poor fit of the continental margins. Geological evidence does not support a high degree of subduction. Owen shows that an exact fit of the various continental fragments together to reform Pangaea, which agrees with geometric and geological matches, is obtained where the value of the earth's curvature is increased to the point at which the diameter of the earth is 80 per cent of the current mean. This corresponds in time to the late Triassic–early Jurassic. This has interesting consequences. Shields (1979, 1983), for example, argues that earth expansion would close the Pacific entirely prior to the Mid-Mesozoic, which means that the Pacific Ocean opened as late as Jurassic times.

All three theories compete over the arrangement of land masses in a supercontinent and details of timing of rifting. As we wrote this monograph, the general consensus of biologists and geologists favoured the Pangaean hypothesis, whereas the Pacific islands, Pacifica and the expanding earth hypotheses are generally looked upon with scepticism, in much the same way as many scientists viewed the Pangaean hypothesis nearly twenty years ago.

Either supporters of the Pacific islands or Pacifica hypothesis are correct, or supporters of the Pangaea hypothesis are correct, or all are correct to some degree, where problems of timing are overcome by an expanding earth model. What can biogeographers do to help resolve the dilemma?

The acceptance of plate tectonics as a geological theory has been regarded as one support of vicariance biogeography: if the continents have drifted apart then so have the plants and animals on them. A particular theory of the sequence of splitting of a supercontinental land mass makes certain predictions regarding the relationships of biotas on the land masses. The Pacifica hypothesis predicts close relationship of biotic elements on allochthonous terrains on either side of the Pacific Ocean (for example, parts of Western North America and Asia); whereas the Pangaean hypothesis predicts close relationship of the biotas on either side of the Atlantic Ocean (for example, eastern North America and Europe). Vicariance and cladistic biogeographers have tested the Pangaean hypothesis by comparing their phylogenies to a sequence of the separation of land masses (e.g. Humphries 1981; Parenti 1981*a*; Rosen 1974). Many groups fit the sequence well, at least in part, for example killifishes and cichlids; others such as midges and beeches do not. In general, pantropical groups fit the sequence of Pangaean break-up rather well, whereas amphitropical groups do not.

Melville (1981, 1982) presented numerous examples of trans-Pacific relationships among plants in support of the Pacifica hypothesis. These included buttercups in New Zealand and South America, carrot-like *Oreomyrrhis* across the southern Pacific and in the Andes from Colombia to northern Argentina and shrubby *Coriaria*, *Gunnera* and the southern speedwells of *Hebe* in Chile and New Zealand. Similarly, Shields (1983) notes that formal trans-Pacific links are many, to support a Jurassic split for the Pacific. These include the Microhylinae, a subfamily of small tree frogs, the mole genus *Urotrichus* in Japan and California, and the snake subfamilies Xenoderminae and Pareinae (Colubridae) in the eastern orient and tropical America. Crayfish distributions and various South-east Asian groups suggest connections of East Asia with North America and South-east Asia with Australia and Madagascar.

Thus, some biogeographers have claimed that the relationships represented in their subgroups support the Pangaean hypothesis, whereas we, like Shields (1983), Melville (1981, 1982), Nur and Ben Avraham (1981), and Carey (1976) make alternative claims that relationships represented in other groups support a Jurassic Pacific opening as in the Pacifica or expanding earth hypotheses. Because we believe there must be one comprehensive theory, rather than three, to explain the world-wide distribution of biotas, we address the problem of historical biogeography at the southern end of the world to analyse further these three hypotheses.

4.4 HISTORICAL BIOGEOGRAPHY OF THE SOUTHERN END OF THE WORLD

Consistent with the foregoing discussions of the Pacifica versus the Pangaea hypothesis is the idea that these are not competing hypotheses at all but are both true, at least in part. The contrast between these competing theories can be seen well in an analysis of the distribution patterns in the southern continents, and their relationship to those of the boreal zone, the problem of amphitropical distributions.

4.4.1 Darlington's view (1965)

Despite the pleas of Croizat (1952, 1958, 1964), van Steenis (1962), and Craw (1982) that ancient geography demands independent interpretations of present day distribution patterns, many present-day biogeographers are traditionalists who simply accommodate Darwinian dispersal hypotheses onto fashionable geological theories rather than attempting to find general explanations for biological and geological data. Raven and Axelrod (1974, p. 539) went as far to say that 'plate tectonic theory does not require any modifications of previously established major principles of evolution'. For example, their view of the southern end of the world is simply a modification of Darlington's (1965) view. By this they mean that plants originated in a 'centre of origin' (west Gondwanaland) and then migrated by overland dispersal routes. For them the austral flora and fauna originated in South America: 'Australasia remained open to immigration from South America by island stepping stones for plants and animals of cool temperate requirements into the mid-Tertiary.' (Raven and Axelrod 1974, p. 635). In other words there is no critical appraisal of plate tectonics by biogeographical patterns. To our minds, Darlington's work sums up best the problems of traditionalist dispersalist biogeography. His view of the distribution of the biota of the southern end of the world was limited by at least two assumptions:

(a) a pre-plate tectonic version of geology, i.e. since the late Cretaceous the southern continents have not been in contact with each other (or connected by land bridges); and

(b) the assumption that a taxonomic group is as old as its earliest fossil representative.

Therefore, because many plant and animal groups are not represented by fossils as old as the late Cretaceous, Darlington concluded that each group found on the southern continents got there by dispersal from its own centre of origin. As distinct from Raven and Axelrod (1974), Darlington admitted that there were conceptual difficulties with postulating dispersal of a variety of organisms across water gaps, but he stated that he was 'forced' (p. 158) to this conclusion by the other evidence of age of groups and prior continental contact. He reviewed several possible mechanisms of dispersal, judging some (such as transport on ice floes) to be nearly impossible, and others (such as transport by strong circumpolar winds) to be quite important in creating southern hemisphere distributions.

By keeping to the two assumptions of age of groups and lack of relatively recent connection of southern continents, Darlington was forced to conclude that dispersal has been the prime factor in creating the distribution of the southern continental biota. Darlington's assumptions precluded a precise examination of organism relationships as well as a search for patterns among the relationships. Also, Darlington's method, and that of fitting a phylogeny to a model of Pangaean break-up, cannot lead to suggestions for reinterpretations of prevailing geological theory because they are wholly dependent upon the prevailing theory. We examine this problem further by concentrating on competing biological and geological theories concerning one of the continents in the southern hemisphere, South America.

4.4.2 The two South Americas

In a study of the historical biogeography of the southern beech genus *Nothofagus*, Humphries (1981) presented the area cladogram of Fig. 4.8 as a summary of the historical relationships of land masses inhabited by austral and boreal groups, based on the cladograms of relationships of many plant and animal taxa, incorporating his data with those compiled by Patterson (1981*a*). Humphries concluded that North America and Europe are sister areas, and that there is a group of austral areas comprising New Zealand, Tasmania, Australia, New Caledonia, and New Guinea. South American taxa either have close relatives in the northern hemisphere, that is in North America or Europe, or they have relatives in the southern hemisphere, in Australia and associated areas, or relatives in other tropical areas, such as tropical Africa.

The area of South America that is included within the austral zone is Patagonia, that is, South America south of approximately 30° south latitude. Most lowland tropical South American taxa have affinities with other tropical groups, or with northern hemisphere biotic elements. Humphries (1981, p. 205) concluded that: 'The two positions for South America . . . are probably due to the fact that it is a huge composite area and should not be treated as a single area of endemism.'

Commenting on the area cladograms for widespread groups presented by Patterson (1981*b*), Parenti (1981*b*) concluded that because South America has affinities with northern hemisphere and southern hemisphere elements, it should be divided into two regions for biogeographical analysis. These regions were referred to as South America 1 and South America 2 (Fig. 4.9); although precisely where the continent should be divided was not specified.

There is a possible alternative explanation for the different relationship patterns for areas of South America. Those organisms showing a close

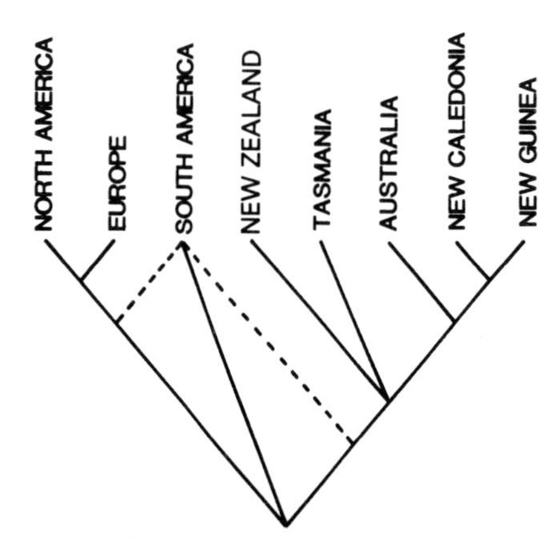

Fig. 4.8. Area cladogram summarizing data for austral, boreal, and pantropical groups, showing the disjunct relationship of the South American continent. (After Humphries 1981.)

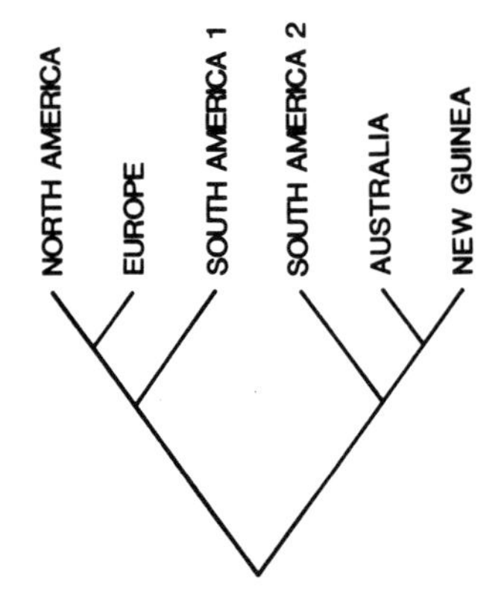

Fig. 4.9. Area cladogram summarizing data of Patterson (1981*a*) for austral, boreal, and pantropical groups, showing the disjunct relationship of the South American continent. (After Parenti 1981*b*.)

relationship to other southern continent taxa may be relatively older groups, with older Pacific and Atlantic north-south connections that were affected differently by past land connections, the more recent South American groups showing relationships to North American and European taxa.

We examine some of the evidence in favour of both of these hypotheses, and then review how cladistic biogeography presents the information derived from a phylogenetic analysis in the form of hypotheses that concur with or refute such conclusions.

4.4.2.1 Patterns of taxa

South American taxa have their closest relatives in one of three areas: (a) the north temperate region (e.g. North America or Europe); (b) the southern cold temperate region (e.g. Australia and other land masses of the austral zone or (c) other tropical regions (e.g. tropical Africa).

Both Patterson (1981*a*) and Humphries (1981) reviewed the data available on relationships of several widespread groups to reach some conclusion concerning the distribution of marsupials (Patterson) and southern beeches (Humphries), two groups with representatives in South America. The data were presented in the form of area cladograms; that is, cladograms of taxa were redrawn with the name of the area inhabited by a taxon replacing the taxon name (see Chapter 2). The area cladograms repeatedly show South America with at least two associations (Fig. 4.10). South American taxa may have as their closest relative a member of a North American-European or African group, rather than an Australian–New Guinean group (Fig 4.10(a)), as in hylid tree frogs, the large flightless ratite birds and galliform (game) birds and xylotine sawflies. Alternatively, the South American taxa are more closely related to taxa in Australia and New Guinea than to taxa in North America or Europe (Fig. 4.10(b)) as in southern beeches, and two subfamilies of chironomid midges. Any cladogram summarizing the data given in Fig. 4.10(a) and (b) must, like Fig. 4.8, show South America in a hybrid position between the austral region (including Australia and New Guinea) and the boreal region (north America and Europe), or as two separate areas, as in Fig. 4.9.

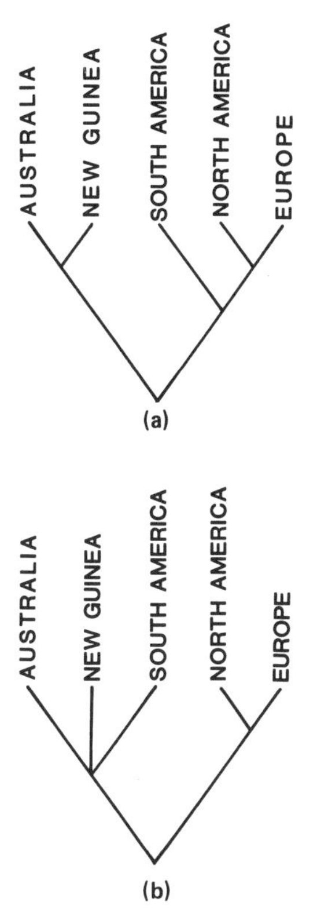

Fig. 4.10. (a) Area cladogram of coincident land masses inhabited by hylid frogs, ratite and galliform birds, and xylotine flies. (After Patterson 1981*a*.)
(b) Area cladogram of coincident land masses inhabited by beeches of the genera *Nothofagus* and *Fagus* and podonomine and diamesine midges. (After Patterson 1981*a*.)

Because all patterns on a worldwide scale represent generalizations from distributions, exceptions to the general pattern of Fig. 4.8 can be found. For example, the South American species of the plant genus *Euphrasia* have their closest relatives in the southern and western Pacific, rather than in North America, where *Euphrasia* species also occur (Du Rietz 1960). The killifish genus *Orestias* is hypothesized to be related more closely to an Anatolian genus than to any South or

North American or African killifish genus (Parenti 1981*a*). These exceptions to the general rule may either represent other general patterns, of different age, or can be explained by events unique to *Euphrasia* or to killifishes (such as independent dispersal). A third explanation is that sometimes exceptions to general rules cease to be exceptions when they are looked at in detail.

The South American taxa that follow the pattern of Fig. 4.10(a) are found either in the tropical lowlands or in the Andean highlands north of 30° South latitude. The South American taxa that follow the pattern of Fig. 4.10(b) are found in South America South of 30° South latitude, in Patagonia. These latter taxa are involved in typical austral zone distributions.

That Patagonia contains a biota distinct from that of tropical lowland South America has been recognized nearly as long as the distinctness of the austral zone has been recognized (see von Ihering 1900, Eigenmann 1909). *Euphrasia* occurs in the western part of Patagonia southwards from approximately 30° latitude. It is not a member of the tropical lowland biota and therefore would not be expected to be closely related to North American taxa, given the general pattern of Fig. 4.8.

The killifish genus *Orestias* is sympatric with midges in the Central Andes. Its sister-group relationship to an Anatolian genus may represent dispersal, widespread extinction, or be part of the general pattern of Fig. 4.8, if the Anatolian biota is part of the boreal zone biota. Answering that question is beyond the scope of our current interests. We mention these examples to point out that exceptions to general rules need to be examined in detail before they can be treated as alternative hypotheses.

Another group that has a distribution pattern incongruent with Fig. 4.8 is the marsupials. Theories relating to their historical distribution were reviewed at length by Patterson (1981*a*). We do not review these again because current phylogenies of marsupials may not be correct (McKenna 1980) and only sound phylogenies can be the bases of biogeographic analysis.

4.4.2.2 Patterns of areas

The distinct biological history of Patagonia (and possibly also of the Central Andes) is established. We may consider this area to be the South America of Fig. 4.9. But what does this biotic distinctness mean in terms of the history of the continent?

A summary of the distribution of taxa inhabiting the southern continents and their boreal relatives can be presented in one cladogram of generalized areal relationships (Fig. 4.11).

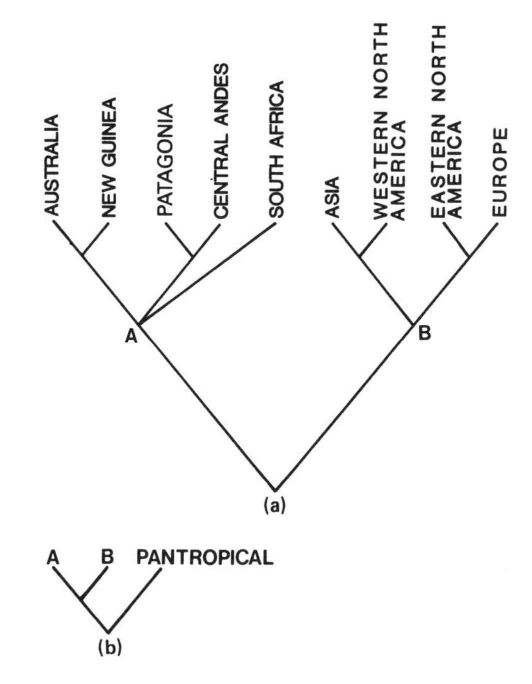

Fig. 4.11. (a) Area cladogram for taxa on the southern continents (A) and their boreal relatives (B). (b) Small cladogram summarizing the relationship of A + B to taxa in the pantropical zone.

Four of its major implications are as follows:

1. Representatives of the Central Andes have their closest relatives in Patagonia, rather than lowland South America.

2. Patagonian taxa, in turn, are members of an austral group related to both South Africa and to the Australian region (including New Guinea, New Caledonia, New Zealand, and Tasmania).

3. The closest relatives of the austral groups are found in the boreal zone. Members of the boreal zone are related in the following pattern: Asia is most closely related to western North America, eastern North America is most closely related to Europe, and these two groups form sister areas.

4. The closest relatives of the austral and boreal groups are found in the pantropical region.

This view of the world as presented in one area cladogram (Fig. 4.11) is derived from relationships among plants and animals. If we believe that the world and its biota evolved together, this summary of area cladograms should be able to tell us something about the historical relationships of land masses.

The relationship of austral and boreal groups to pantropical groups also has implications for the type of model of earth history one believes is supported by such patterns of biotic relationships. For example, even though the relationships of the spiny-finned percichthyid fishes, a typical amphitropical group, have not been worked out by fish systematists, one prediction of the general pattern is that the northern hemisphere taxa form a monophyletic group that is the sister group of a monophyletic southern hemisphere taxon. This would require trans-Pacific relationships in both the northern and southern hemispheres, and imply the existence of a Pacific continent. On the other hand it would also require transatlantic relationships in both hemispheres, and support the former existence of an Atlantic continent. No one theory of continental splitting, Pacifica, expanding earth or Pangaea, agrees in total with what the relationships of taxa suggest to us about past land connections. However, we can also derive an independent theory of the relationship of land masses from geological data. On some level, area cladograms derived by systematists should be congruent with relationships of the land masses derived by geologists.

Can geological information be used separately to derive a theory of relationships of the southern hemisphere land masses? We continue to look at one continent, South America, in detail.

The central and southern Andes are characterized by rock (estimated to be as old as the Jurassic) lying on top of continental South America. These regions were termed 'Peruvian Pacifica' and 'Magellanian Pacifica' by the botanist Melville (1981) (Fig. 4.12). But why 'Pacifica'?

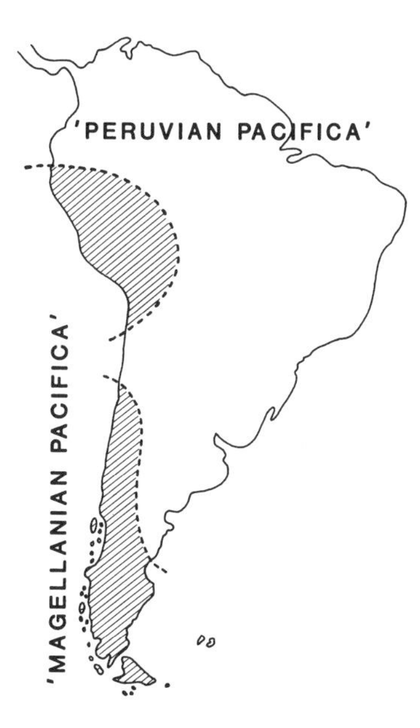

Fig. 4.12. Map showing two regions of South America, Peruvian, and Magellanian Pacifica, postulated by Melville (1981) to be derived from an ancient Pacific continent.

In a recent review of the geologic history of the Andes, the geologist Werner Zeil (1979) stated that the Andes of Colombia, Ecuador, and Venezuela fit the plate tectonic model for the history of South America, and correlate with classical hypotheses of Andean mountain building by subduction of a Pacific plate. The Andes of Peru, Bolivia, Chile, and Argentina, Zeil argued, do not correspond to the plate tectonic model. To quote Zeil: 'The wealth of morphological features is a reflection of the geologically different elements which make of (sic) the coastal range of the Andes'. He characterized the Andes in Colombia and Ecuador as remnants of a late Mesozoic island arc or oceanic crust welded to the central and eastern Cordilleras of the continental Andes.

The central and southern parts of the range (Peru, Chile) are continental crust without typical oceanic material but probably consist of a large number of different tectonic segments.

Other geologists have opposed this explanation, or at least debated it. We emphasize here that geologists disagree on the particular facts relating to an area's history in the same way as systematists disagree on relationships among taxa. Important questions, such as whether or not South America fits the plate tectonic model, are still unanswered. James (1973), in his hypothesis for the evolution of the Central Andes, comments on this Jurassic Peruvian rock, stating that: 'the rocks could be part of a Palezoic microcontinent or peninsula that lay to the west of the South American coastline; or they could be sialic flotsam swept into and plastered to the edge of South America, buoyant debris scraped from the top of the oceanic plate as it dived down at the trench.' (James 1973, p. 65.)

Nur and Ben Avraham (1981) had little doubt that the exotic terranes found around the Pacific margin, and including at least in part the central and southern Andes, are part of an ancient Pacific continent they termed Pacifica. Alternatively other geologists, (Batten and Schweickert 1981) as well as palaeontologists (Tedford 1981; Tozer 1982) had little doubt that these terranes had nothing to do with an ancient Pacific continent. Jones *et al*. (1982) reviewed the history of trans-Pacific land masses, particularly North America and Asia, and concluded that western North America has grown by collisions with small land masses, some of which originated far to the west of the present continent.

As biologists, we are unable to evaluate the different theories proposed by geologists for the history of these land masses. To derive a cladogram from geological data including the central and southern Andes, certain questions must be answered: are the rocks of the Peruvian and Magellanian Pacifica of the same origin? That is, are their similarities derived similarities that would indicate shared geological history? If so, to what land mass are they next most closely related; that is, what land mass shares unique properties with the Peruvian and Magellanian Pacific? And is the land mass continental or extra continental? Without answers to these questions biologists with good evidence can only suggest that their cladograms of taxa accurately reflect earth history. Biological evidence indicated continental drift well in advance of its eventual acceptance. Geologists should re-examine their theories to see if their hypotheses really are good fits to the geological facts.

4.4.2.3 Geology or age?

The separate biotic associations of tropical and temperate southern South America suggest at least two conclusions: (a) South America is a continent of hybrid origin; or (b) there is an older and younger component to these patterns, that is the association of part of South America with the austral zone represents a distribution pattern of older groups that has since been overlain by younger groups that have affinities with other tropical representatives.

The biological data represented by area cladograms (Figs 4.8-4.11) present a considerable challenge to geologists to address the problems that still prevent a full understanding of the geological history of South America, as we have discussed above.

Can the relative ages of groups aid our understanding of the origin of the South American dichotomy? It is true that many pantropical groups fit a model for the break-up of Pangaea rather well. These groups also tend to be relatively young in terms of geological age. For example, killifishes, which fit the Pangaean model to a great extent, are represented by fossils as old as the Oligocene (Parenti 1981*a*). Amphitropical groups tend to be relatively older; beeches (Humphries 1981), beetles (Crowson 1980) and midges (Brundin 1966) are at least as old as the Jurassic. The fact that these groups are at least as old as the Jurassic suggests that the pattern of Fig. 4.11 may be this age. This follows from the idea that if we have a pattern, all taxa sharing the pattern came to occupy the same areas at the same time in response to the same geological events. If this is not true, then any other taxon sharing the pattern and represented by more recent oldest fossils (such as *Nothofagus*) would have to have formed the same distribution pattern at a later date. Adherence to the principle of parsimony precludes this conclusion. That is the nature of a pattern in cladistic biogeography.

Now, with available estimates of ages of groups, and hence of patterns, we conclude that the amphitropical distributions of beetles, midges, beeches, *Euphrasia*, lampreys, and percichthyids, among other groups, are older than the pantropical distribution (shared by cichlids, killifishes, and ebonies, among others). The older amphitropical distribution will remain our conclusion until an even older fossil is found in the pantropical zone that is a member of a group conforming to a Pangaean break-up pattern.

Geologists could aid our understanding of the problem by suggesting relative ages of continental separation or movement of exotic terranes. This has been done for the Pangaea model in its many formulations (see Humphries 1981), which is estimated to have started breaking up in the Jurassic. Shields (1979, 1983) estimates the initial opening of the Pacific in an expanding earth model as Jurassic (Shields 1979). Thus, geological estimates really do not help very much in deciding which pattern of break-up occurred first, or if they were concurrent.

4.4.3 A composite New Zealand

Cladistic biogeography then is not concerned solely with determining the sequences of fragmenting biotas in terms of plate tectonics; it is concerned ultimately with an explanation of the composite areas. As originally demonstrated by Croizat (1952, 1958) and elaborated by Craw (1979), 'track' analyses show that in addition to South America, areas such as New Zealand have complex biogeographic patterns, showing a variety of overlapping relationships with neighbouring areas. Figure 4.13 shows three separate tracks for New Zealand, one shared with South

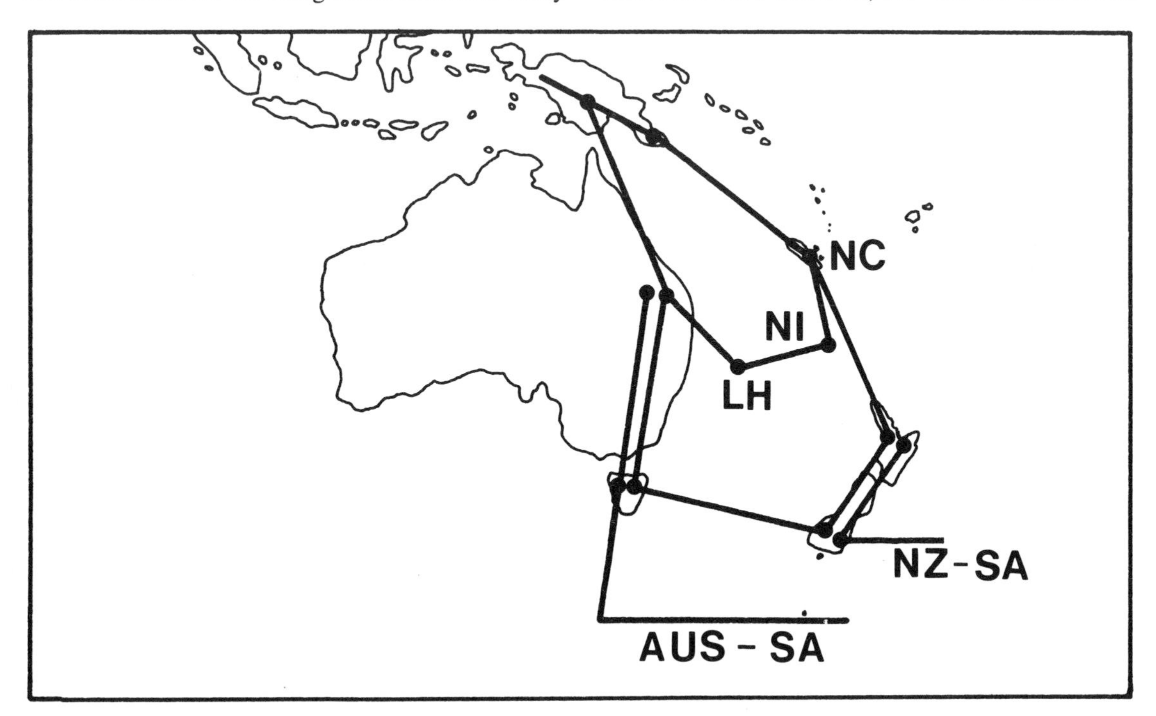

Fig. 4.13. Generalized tracks in and around the Tasman Sea. NC, New Caledonia; NI, Norfolk Island; LH, Lord Howe Island; NZ, New Zealand; SA, South America; AUS, Australia.

America, one with Tasmania, Eastern Australia and New Guinea, and a third with New Caledonia. One cladogram showing a model for Pangaea (Humphries 1981, Fig. 12a) can be compared to two cladograms based on geological data derived by Craw (1982, Fig. 4.14(b), (c)) which emphasize the hybrid composition of New Zealand. Craw's models have support since Howell (1980) proposed that New Zealand is composed of four exotic terranes—microplate accretions along the Gondwanian margin—two related to Australia and Pacifica and two of uncertain origin.

How do the cladograms based on geological data compare with particular groups? Figure 4.15(a) and (b) gives two simplified patterns for three areas for five widely different groups. The first pattern is formed on the basis of two plant groups: *Nothofagus* (Humphries 1981) and a combined cladogram for three genera of the Persooninae (Weston, in Craw 1982). The second pattern is a common one, as shown by a combined cladogram of four genera of the Gesneriaceae (African violets; Humphries 1981), a genus of caddis flies (*Hydrobiosella*; Ross 1956) and a group of platycercine parrots (Craw 1982). The two patterns, if viewed superficially, are incongruent. So what does this mean biogeographically? If continental fragmentation is considered, treating New Zealand as a single area would suggest that *Nothofagus* and the Persooniinae were uninformative. That is the cladograms for three groups, *Hydrobiosella*, Gesneriaceae, and parrots agree with a continental break-up pattern but the cladograms for *Nothofagus* and the Persooniinae are incongruent with such a pattern. However, such a view is geologically naïve. If New Caledonia consists of a minimum of two continental fragments, one of which separated from Australia during the Jurassic, and the other which separated as a result of the opening of the Tasman sea during the Upper Cretaceous-Palaeocene, then groups associated with the older and younger sequences would naturally be incongruent (Craw 1982). As Craw noted, the *Nothofagus*/Persooniinae pattern (Fig. 4.15(a)) would be associated with the Jurassic fragment, while the second pattern (Fig. 4.15(b)) is due to rafting out of a biota in the north-east Australia-New Caledonia/New Zealand track.

4.5 A NEW VIEW OF THE WORLD

Darlington (1965) was 'forced' to conclude that climate and dispersal were the primary factors in establishing amphitropical patterns as well as the similarity of biotas on the southern continents because he made two assumptions:

(a) the southern continents were relatively stable during the period of evolution of taxa on the land masses, precluding the existence of a former large supercontinent in the southern hemisphere; and

(b) the taxa occupying these land masses are too young to have been affected by any postulated land movement.

However, we have disregarded these assumptions and have seen that an alternative view of the world is possible. Our view suggests a return, in part, to the ideas of Humboldt and Hooker on the importance of former land connections in the southern hemisphere.

4.5.1 Pacifica and amphitropical distributions

In the previous discussions in this chapter, we concluded that

(i) an explanation of the distribution of taxa on the southern continents is related to amphitropical distributions;

(ii) the close relationship of austral groups to boreal groups, rather than to pantropical groups, suggests a former connection between the austral and boreal zones; and

(iii) certain aspects of both expanding earth and Pacifica hypotheses coincide with relationships among amphitropical groups.

Biogeographers are still left with an intriguing problem: how does one explain amphitropical distributions?

If one believes the relationships of taxa can tell us something about the form of past land connections, then the following model is a plausible

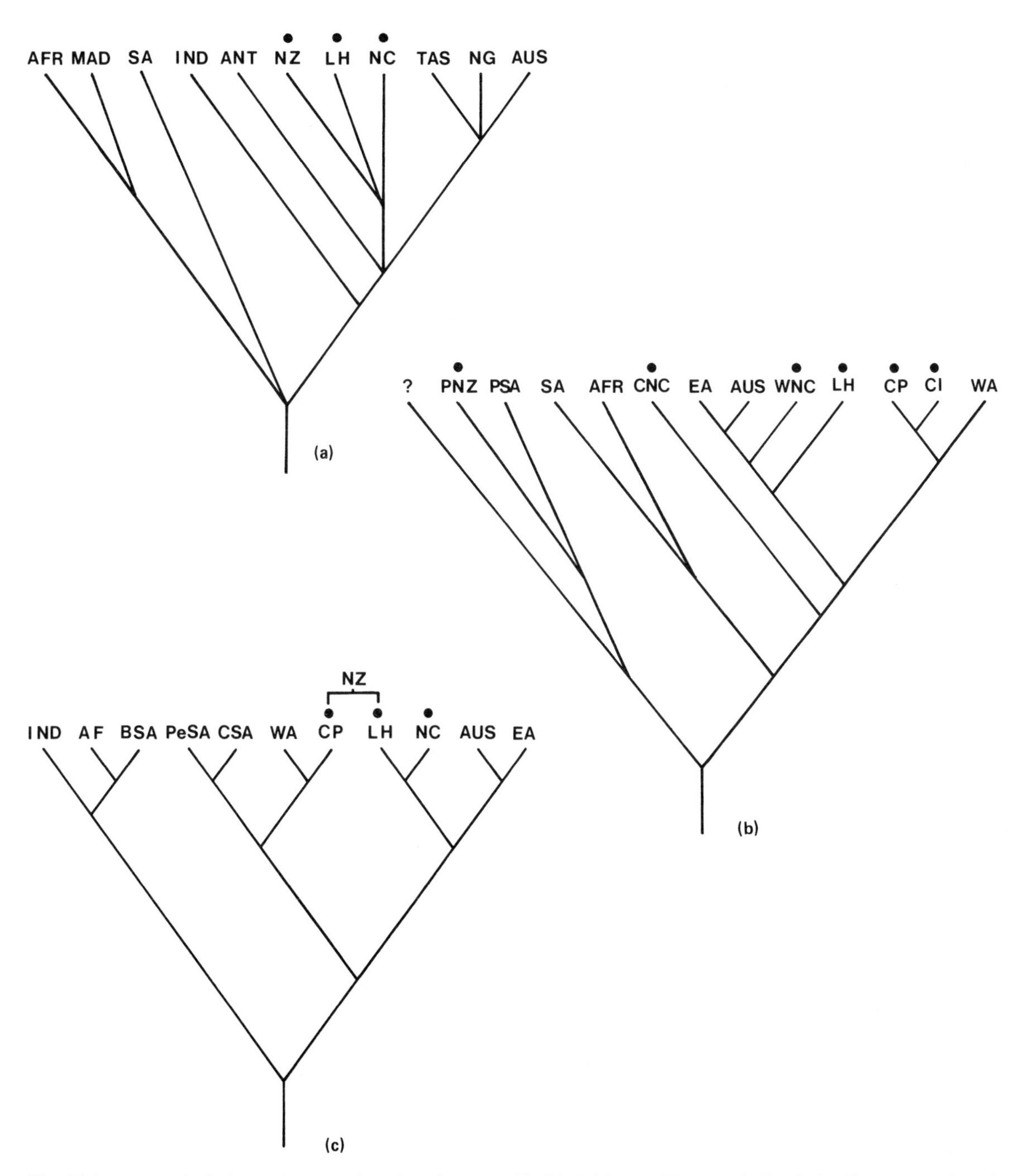

Fig. 4.14. (a) A geological area cladogram based on the maps of Smith, Briden, and Drewry (1973). (b) Pacifica construction of Nur and Ben Avraham (1981) and Kamp (1980). (c) Based on Pacifica reconstruction of Melville (1981). Abbreviations: Afr., Africa; NZ, New Zealand; LH, Lord Howe Island; NI, Norfolk Island; NC, New Caledonia; Tas, Tasmania; NG, New Guinea; Aus, Australia; CNC, Central Chain New Caldeonia; EA, East Antarctica; WNC, west coast New Caledonia; CP, Campbell Plateau; CI, Chatham Islands; WA, west Antarctica; BSA, Brazilian South America; PeSA, Peruvian South America; CSA, Chilean (Magellanic) South America. ●—Tasmantis.

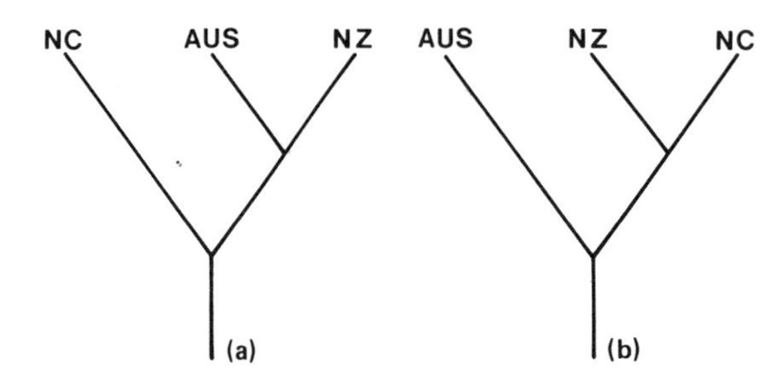

Fig. 4.15. Reduced area cladograms. (a) *Nothofagus*; Persooniinae (*Persoonia*—Aus; *Toronia*—NZ; *Garnieria*—NC). (b) *Hydrobiosella*; Gesneriaceae (Fieldia—Aus; *Coranthera*, *Depanthes*—NC; *Rhabdothamus*—NZ) Platycercine parrots.

explanation for amphitropical distributions (Fig. 4.16).

Austral and boreal groups were at one time adjacent, and they in turn were adjacent to a tropical zone. Movement of the austral zone to the opposite side of the tropical zone would present a world called Pangaea, ready to break up at the end of the Triassic to form patterns of relationship of land and taxa consistent with our present hypotheses concerning relationships. Trans-Pacific as well as transatlantic patterns would accompany the opening of the Pacific and Atlantic oceans; and these would be austral and boreal.

We term the hypothetical continent in which austral and boreal zones are adjacent 'pre-Pangaea'. We could just as well have called it Pacifica, because like all continental masses proposed under that name, it explicitly requires the juxtapositions of current trans-Pacific areas.

4.5.2 Pacifica versus Pangaea: geology or age?

The model of Fig. 4.16 is one of many that could be constructed to describe a pre-Pangaean configuration of land masses. Others have been proposed but with earlier dates, primarily by geologists (see, for example, Bambach, Scotese, and Ziegler 1980). Although we cannot give a geological mechanism, this model does have an important relative correlation with age. Basic amphitropical distributions are older than pantropical distributions. We hesitate to give an absolute estimate of the timing of the events in this model, primarily because both geologists and biologists have presented the break-up of a Pacific continent in contrast to the break up of Pangaea, rather than considering the possibility that if some sort of Pacifica existed, it existed much earlier than Pangaea, and therefore, would be of an older geological period.

4.6 CONCLUSIONS

In this chapter, we have attempted to provide one hypothesis of explanation for both amphitropical and austral distributions, coordinate problems that have captured the interest of biogeographers for over a century. This investigation has, by necessity, expanded to one in which we examine

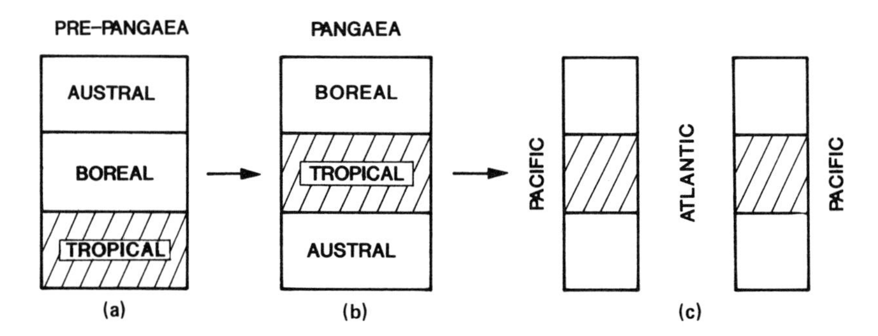

Fig. 4.16. Schematic diagram showing hypothetical relationships of austral, boreal and pantropical zones. (a) Pre-Pangaean configuration with austral and boreal zones adjacent. (b) Pangaean configuration with austral zone having moved to opposite side of tropical zone. (c) Rifting to form Atlantic Ocean (earlier division of Pangaean continent into Laurasia and Gondwanaland not shown).

explanations proposed for the related problem of disjunct distributions of South American and New Zealand taxa.

As cladistic biogeographers, our primary concern is finding a historical explanation for a global distribution. We looked first for patterns of relationships among the austral and amphitropical groups for which reliable cladograms of relationships had been proposed. Our task was made easier by finding a single explanation for the cladograms of two subfamilies of midges and the beeches of the genera *Fagus* and *Nothofagus* because the midges (Brundin 1966) and beeches (Humphries 1981) have figured prominently in many studies of austral and amphitropical biogeography.

The finding of a pattern (Fig. 4.11) corroborated in whole or in part by numerous other groups of plants and animals invites a general explanation for the distribution of austral zone and amphitropical distributions as well as for the hybrid composition of South America and the complex composition of New Zealand. In a search for an explanation, we have abandoned several restrictions placed on many previous biogeographic studies. These restrictions are:

(a) that the prevailing geological theory (in this case the former existence and break-up of the supercontinent Pangaea) should be the only framework within which our pattern is interpreted; and

(b) that a group is estimated to be only as old as its oldest fossil.

Our pattern for disjunct biotic associations within South America, which includes the hypothesis of close relationship of the central Andes to Patagonia, and hence inclusion of the central Andes in the problem of defining austral and amphitropical distribution patterns, correlates in a striking way with hypotheses of some geologists (Carey 1976; Shields 1979, 1983; Nur and Ben Avraham 1981) as well as the biologists Croizat 1964; Nelson and Platnick 1981; Melville 1981) for an extra-continental origin of these areas of South America.

The incorporation of a pattern of area relationships of the austral and boreal zones (including an implied relationship of pantropical taxa) with the Pacifica, expanding earth and Pangaean hypotheses, poses these separate theories of taxa and area relationship not as independent theories but as part of an integrated whole. It is not, however, the final answer. To some it will not even be a plausible answer, for as Darlington said, 'No biogeographer would seriously suggest ... that existing north and south temperate areas once formed a single land mass. ...' Were it the final answer, biogeography would be over, for we would have explained the world, and we have not done that.

We regard the hypothesis of Fig. 4.11 as just one of many that could today be put forward as an explanation of amphitropical patterns. The significance of cladistic biogeography in establishing this pattern is that it is not based on one group of organisms, but on a series of unrelated taxa.

We believe that there is just one underlying comprehensive theory of geological evolution that when proposed will explain most large-scale distribution patterns. That comprehensive theory will be based on an exhaustive set of both biological and geological cladograms. When cladograms of relationship of all taxa, worldwide, are known, the pattern we have presented in Fig. 4.11 could turn out to be the most unusual among all groups of taxa and explicable only by the dispersal of individual taxa. But for now it is our only comprehensive hypothesis of amphitropical distributions and relationships, incorporating the pioneering cladistic data presented for midges and beeches. As such it remains to be corroborated or refuted by cladistic analyses of many additional amphitropical groups.

BIBLIOGRAPHY

Adams, C. C. (1902). South eastern United States as a center of geographical distribution of fauna and flora. *Biological Bulletin of the Marine Biology Laboratory, Woods Hole* **3**, 115-31.

Bailey, I. W. (1949). Origin of the angiosperms: need for a broadened outlook. *Journal of the Arnold Arboretum, Harvard University* **30**, 64-70.

Ball, I. R. (1976). Nature and formulation of biogeographic hypotheses. *Systematic Zoology* **24**, 407-430.

Bambach, R. K., Scotese, C. R., and Ziegler, A. F. (1980). Before Pangaea: the geographies of the Paleozoic world. *American Scientist* **68**, 26-38.

Batten, R. L. and Schweikert, R. A. (1981). (Discussion of) Lost Pacifica continent: a mobilistic speculation. In *Vicariance biogeography: a critique*. (eds G. Nelson and D. E. Rosen). pp. 359-366. Columbia University Press, New York.

Beck, C. B. (1976). Origin and early evolution of angiosperms: a perspective. In *Origin and Early Evolution of Angiosperms* (ed. C. B. Beck) pp. 1-10. Columbia University Press, New York .

Berra, T. M. (1981). *An atlas of distribution of the freshwater fish families of the world*. University of Nebraska Press, Lincoln, NB.

Bremer, K. and Wanntorp, H. E. (1981). The cladistic approach to plant classification. In *Advances in cladistics: proceedings of the first meeting of the Willi Hennig Society* (eds V. A. Funk and D. R. Brooks) pp. 87-94. New Botanical Garden, New York.

Brundin, L. (1966). Transantarctic relationships and their significance as evidenced by midges. *Kungliga Svenska Vetenskapsakademiens Handlinger*, Series 4, **11**, 1-472.

Brundin, L. (1972*a*). Evolution, causal biology, and classification. *Zoologica Scripta* **1**, 107-120.

Brundin, L. (1972*b*). Phylogenetics and biogeography. *Systematic Zoology* **21**, 69-79.

Brundin, L. (1981). Croizat's panbiogeography versus phylogenetic biogeography. In *Vicariance biogeography: a critique* (eds G. Nelson and D. E. Rosen) pp. 94-158. Columbia University Press, New York.

Buffon, G. L. L. Compte de (1776). *Histoire naturelle generale et particuliere*, Supplement III *Servant de Suite à l'histoire des animaux Quadrupèdes*. A Paris de de l'Imprimerie Royale, pp. 330 + xxj.

Bush, G. L. (1975). Modes of animal speciation. *Annual Review of Ecology and Systematics* **6**, 339-364.

Cain, S. A. (1944). *Foundations of plant geography*. Harper and Row, New York.

Camin, J. and Sokal, R. R. (1965). A method for deducing branching sequences in phylogeny. *Evolution* **19**, 311-326.

Candolle, A. P. de (1820). Essai elementaire de geographie botanique. In *Dictionaire de sciences naturelles*, Vol. 18. Flevrault, Strasbourg and Paris.

Carey, S. W. (1976). *The expanding earth*. Elsevier, Amsterdam.

Corner, E. J. H. (1959). Panbiogeography (review of Croizat 1958). *New Phytologist* **58**, 237-38.

Craw, R. (1982). Phylogenetics, areas, geology and the biogeography of Croizat: a radical view. *Systematic Zoology* **31**, 304-316.

Croizat, L. (1952). *Manual of phytogeography*. W. Junk, The Hague.

Croizat, L. (1958). *Panbiogeography*. Published by the author, Caracas.

Croizat, L. (1964). *Space, time and form, the biological synthesis*. Published by the author, Caracas.

Croizat, L. (1982). Vicariance, vicariism, panbiogeography, 'vicariance biogeography', etc. A clarification. *Systematic Zoology* **31**, 291-304.

Croizat, L., Nelson, G., and Rosen, D. E. (1974). Centers of origin and related concepts. *Systematic Zoology* **23**, 265-287.

Crowson, R. A. (1980). On amphipolar distribution patterns in some climate groups of Coleoptera. *Entomologia Generalis* **6**, 281-292.

Darlington, P. J. Jr. (1957). *Zoogeography: the geographical distribution of animals*. John Wiley & Sons, New York.

Darlington, P. J. Jr. (1965). *Biogeography of the southern end of the world*. Harvard University Press, Cambridge, MA.

Darwin, C. (1859). *On the origin of species by means of*

natural selection, or the preservation of favoured races in the struggle for life. John Murray, London.

Diamond, J. (1982). The biogeography of the Pacific Basin. *Nature* **298**, 604-605.

Diels, L. (1908). *Pflanzengeographie*. Goschen'sche Verlagshandlung, Berlin.

Dupuis, C. (1978). Systematique phylogenetique de W. Hennig (historique, discussion, choix de references). *Cahiers des Naturalistes* **34**, 1-69.

Du Rietz, G. E. (1940). Problems of bipolar plant distribution. *Acta Phytogeographica Suecica* **13**, 215-282.

Du Rietz, G. E. (1960). Remarks on the botany of the southern cold temperate zone. In A discussion on the biology of the cold temperate zone (ed. C. F. A. Pantin) pp. 500-507. *Proceedings of the Royal Society* (London), **B152**, 429-682.

Eigenmann, C. H. (1909). The freshwater fishes of Patagonia and an examination of the Archiplata–Archhelenis theory. Reports of the Princeton University Expedition to Patagonia, 1896-1899. *Zoology* **3**, 225-374.

Eldredge, N. and Cracraft, J. (1980). *Phylogenetic patterns and the evolutionary process: method and theory in comparative biology*. Columbia University Press, New York.

Endler, J. A. (1982). Problems in distinguishing historical from ecological factors in biogeography. *American Zoologist* **22**, 441-452.

Farris, J. S. (1970). Methods for computing Wagner trees. *Systematic Zoology* **19**, 83-92.

Flenley, J. R. (1979). *The equatorial rain forest: a geological history*. Butterworths, London.

Forbes, E. (1846). On the connection between the distribution of the existing fauna and flora of the British Isles and the geological changes which have affected their area, especially during the epoch of the Northern Drift. *Memoirs of the Geological Survey of Great Britain* **1**, 336-432 (2 maps).

Forey, P. L. (1981). Biogeography. In *Chance, change and challenge*, Vol. 2, *The evolving biosphere* (ed. P. L. Forey) pp. 241-245. British Museum (Natural History) and Cambridge University Press, London and Cambridge.

George, W. (1962). *Animal geography*. Heinemann, London.

Good, R. (1964). *The geography of flowering plants*. Longman, London.

Good, R. (1974). *The geography of flowering plants* (4th edn). Longman, London.

Hallam, A. (1973). *A revolution in the earth sciences; from continental drift to plate tectonics*. Clarendon Press, Oxford.

Hammond, P. (1975). The phylogeny of a remarkable new genus and species of gymnusine staphyl (Coleoptera) from the Auckland Islands. *Journal of Entomology* **B44**, 153-173.

Heer, O. (1868, 1878). *Die fossile flora der Polarlander: flora fossilis Arctica*. Vol. 1 (1868). F. Schutthess, Zurich. Vol. V: J. Wurster and Co., Zurich.

Hennig, W. (1950). *Grundzüge einer Theorie der phylogenetischen Systematik*. Deutscher Zentralverlag, Berlin.

Hennig, W. (1965). Phylogenetic systematics. *Annual Review of Entomology* **10**, 97-116.

Hennig, W. (1966). *Phylogenetic systematics* University of Illinois Press, Urbana, IL.

Hill, C. R. and Crane, P. R. (1982). Evolutionary cladistics and the origin of Angiosperms. In *Problems of phylogenetic reconstruction* (ed. K. A. Joysey and E. A. Friday) pp. 269-361. Academic Press, London.

Hooker, J. D. (1853). *The botany of the Antarctic voyage of H.M. discovery ships Erebus and Terror in the years 1839–1843*. Vol. II, *Flora Novae-Zelandiae*, Part 1, Flowering plants. Lovell Reeve, London.

Hooker, J. D. (1860). *Botany of the Antarctic voyage of H.M. discovery ships Erebus and Terror in the years 1839–1843*. Vol. III, *Flora Tasmaniae*. Lovell Reeve, London.

Howarth, M. K. (1981). Palaeogeography of the Mesozoic. In *Chance, change and challenge* Vol. 1, *The evolving earth* (ed. L. R. M. Cocks) pp. 197-220. British Museum (Natural History) and Cambridge University Press, London and Cambridge.

Howell, D. G. (1980). Mesozoic accretion of exotic terranes along the New Zealand segment of Gondwanaland. *Geology* **8**, 487-491.

Hubbs, C L. (1952). Antitropical distribution of fishes and other organisms. In Meteorology and oceanography, *Proceedings of the seventh Pacific Scientific Congress, Pacific Science Association* **3**, 324-329.

Hudson, W. H. (1892). *The naturalist in La Plata*. Chapman and Hall, London.

Hughes, N. F., Drewry, G. E., and Laing, J. F. (1979). Barremian earliest angiosperm pollen. *Palaeontology* **22**, 513-535.

Humboldt, F. A. von and Bonpland, A. J. A. (1805). *Essai sur la geographie des plantes*, Levrault, Schoell & Compagnie, Paris.

Humphries, C. J. (1981). Biogeographical methods and

the southern beeches (Facaceae: *Nothofagus*). In *Advances in cladistics: proceedings of the first meeting of the Willi Hennig Society* (eds. V. A. Funk and D. R. Brooks) pp. 177-207. The New York Botanical Garden, New York.

Ihering, H. von (1900). The history of the Neotropical region. *Science* **12**, 857-864.

James, D. E. (1973). The evolution of the Andes. *Scientific American* **229**, 60-69.

Jones, D. L., Cox, A., Coney, P., and Beck, M. (1982). The growth of western North America. *Scientific American* **247**, 50-64.

Kamp, P. J. J. (1980). Pacifica and New Zealand, proposed eastern elements in Gondwanaland's history. *Nature* **288**, 659-64.

Keast, A., Erk, F. C., and Glass, B. eds (1972). *Evolution, mammals and southern continents*. State University of New York Press, Albany, NY.

Kluge, A. G. and Farris, J. S. (1969). Quantitative phyletics and the evolution of anurans. *Systematic Zoology* **18**, 1-32.

Koopman, K. F. (1981). Discussion (of T. C. Erwin, Taxon Pulses, Vicariance and Dispersal). In *Vicariance biogeography: a critique* (eds G. Nelson and D. E. Rosen) pp. 184-187. Columbia University Press, New York.

Kozo-Poljanski, B. M. (1922). *An introduction to the phylogenetic systematics of the higher Plants*. Voronezh.

Linnaeus, C. (1781). *Selected dissertations from the Amoenitates Academicae*. Robeson and Robeson, London.

McAlpine, J. F., Peterson, B. V., Shewell, G. E., Teskey, H. J., Vockeroth, J. R., and Wood, D. M. (1981). *Manual of Nearctic Diptera*, Vol. 1. Biosystematics research Institute, Ottawa.

MacArthur, R. H. and Wilson, E. O. (1967). *The theory of island biogeography*. Princeton University Press, Princeton, NJ.

McDowall, R. M. (1964). The affinities and derivation of the New Zealand fresh-water fish fauna. *Tuatara* **12**, 59-67.

McKenna, M. C. (1980). Early history and biogeography of South America's extinct land mammals. In *Evolutionary biology of the New World monkeys and continental drift*. (eds L. R. Ciochon and A. B. Chiarelli), pp. 43-77. Plenum Press, New York.

Malfait, B. T. and Dinkelman, M. G. (1972). Circum-Caribbean tectonic and igneous activity and the evolution of the Caribbean plate. *Geological Society of America Bulletin* **83**, 251-272.

Marshall, L. G. (1979). A model for paleobiogeography of South American cricetine rodents. *Paleobiology* **5**, 126-132.

Marshall, L. G., Butler, R. F., Drake, R. E., Curtis, G. H., and Tedford, R. H. (1979). Calibration of the Great American Interchange. *Science* **204**, 272-279.

Marshall, L. G. and Hecht, M. K. (1978). Mammalian faunal dynamics and the Great American Interchange: an alternative interpretation. *Paleobiology* **4**, 203-206.

Matthew, W. D. (1915). Climate and evolution. *Annals of the New York Academy of Sciences* **24**, 171-318.

Matthew, W. D. (1918). Affinities and origin of the Antillean mammals. *Geological Society of America Bulletin* **29**, 657-666.

Melville, R. (1981). Vicarious plant distributions and paleogeography of the Pacific region. In *Vicariance biogeography: a critique* (eds G. Nelson and D. E. Rosen) pp. 238-274. Columbia University Press, New York.

Melville, R. (1982). The biogeography of *Nothofagus* and *Trigonobalanus* and the origin of the Fagaceae. *Botanical Journal of the Linnaean Society* **85**, 75-88.

Nelson, G. (1978*a*). Ontogeny, phylogeny and the biogenetic law. *Systematic Zoology* **27**, 324-345.

Nelson, G. (1978*b*). From Candolle to Croizat: comments on the history of biogeography. *Journal of the History of Biology* **11**, 269-305.

Nelson, G. (1982). Cladistique et biogeographie. *Compte Rendus de la Société de Biogeographie* **58**, 75-94.

Nelson, G. and Platnick, N. I. (1980). A vicariance approach to historical biogeography. *Bioscience* **30**, 339-343.

Nelson, G. and Platnick, N. I. (1981). *Systematics and biogeography; cladistics and vicariance*. Columbia University Press, New York.

Novacek, M. J. and Marshall, L. G. (1976). Early biogeographic history of Ostariophysan fishes. *Copeia* **1**, 1-12.

Nur, A. and Ben Avraham, Z. (1981). Lost Pacifica continent: a mobilistic speculation. In *Vicariance biogeography: a critique*. (eds G. Nelson and D. E. Rosen) pp. 341-358. Columbia University Press, New York.

Owen, H. G. (1976). Continental displacement and

expansion of the Earth during the Mesozoic and Cenozoic. *Philosophical Transactions of the Royal* Society **281**, 223-290.

Parenti, L. R. (1980). A phylogenetic analysis of the land plants. *Biological Journal of the Linnaean Society* **13**, 225-242.

Parenti, L. R. (1981*a*). A phylogenetic and biogeographic analysis of cyprinodontiform fishes (Teleostei, Atherinomorpha). *Bulletin of the American Museum of Natural History* **168**, 335-557.

Parenti, L. R. (1981*b*). Discussion (of C. Patterson, Methods of paleobiogeography). In *Vicariance biogeography: a critique*. (eds G. Nelson and D. E. Rosen) pp. 490-497. Columbia University Press, New York.

Patterson, C. (1980). Cladistics. *Biologist* **27**, 234-240.

Patterson, C. (1981*a*). Methods of paleobiogeography. In *Vicariance biogeography: a critique* (eds G. Nelson and D. E. Rosen) pp. 446-489. Columbia University Press, New York.

Patterson, C. (1981*b*). The development of the North American fish fauna—a problem of historical biogeography. In *Chance, change and challenge*, Vol. 2, *The evolving biosphere* (ed. P. L. Forey) pp. 265-281. British Museum (Natural History) and Cambridge University Press, London and Cambridge.

Patterson, C. (1982*a*). Cladistics and classification. *New Scientist* **94**, 303-306.

Patterson, C. (1982*b*). Morphological characters and homology. In *Problems of phylogenetic reconstruction* (eds K. A. Joysey and A. E. Friday). Systematics Association special volume No. 21, pp. 21-74. Academic Press, London.

Peake, J. F. (1982). The land snails of islands—a dispersalist's viewpoint. In *Chance, change and challenge*, Vol. 2, *The evolving biosphere* (ed. P. L. Forey) pp. 247-263. British Museum (Natural History) and Cambridge Uondon.

Peake, J. F. (1982). The land snails of islands—a dispersalist's viewpoint. In *Chance, change and challenge*, Vol. 2, *The evolving biosphere* (ed. P. L. Forey) pp. 247-263. British Museum (Natural History) and Cambridge University Press, London and Cambridge.

Platnick, N. I. (1979). Philosophy and the transformation of cladistics. *Systematic Zoology* **28**, 537-546.

Platnick, N. I. (1981). Widespread taxa and biogeographic congruence. In *Advances in cladistics: proceedings of the first meeting of the Willi Hennig Society* (eds V. A. Funk and D. R. Brooks) pp. 223-227. The New York Botanical Garden, New York.

Platnick, N. I. and Nelson, G. (1978). A method of analysis for historical biogeography. *Systematic Zoology* **27**, 1-16.

Raven, P. H. and Axelrod, D. I. (1972). Plate tectonics and Australasian paleobiogeography. The complex biogeographic relations of the region reflect its geologic history. *Science* **176**, 1379-1386.

Raven, P. H. and Axelrod, D. I. (1974). Angiosperm biogeography and past continental movements. *Annals of the Missouri Botanical Garden* **61**, 539-673.

Rosen, D. E. (1974). The phylogeny and zoogeography of salmoniform fishes and the relationships of *Lepidogalaxias salamandroides*. Bulletin of the American Museum of Natural History **153**, 265-326.

Rosen, D. E. (1976). A vicariance model of Caribbean biogeography. *Systematic Zoology* **24**, 431-464.

Rosen, D. E. (1978). Vicariant patterns and historical explanation in biogeography. *Syst. Zool.* **27**, 159-188.

Rosen, D. E. (1979). Fishes from the uplands and intermontane basins of Guatemala: revisionary studies and comparative geography. *Bulletin of the American Museum of Natural History* **162**, 267-376.

Ross, H. H. (1950). *A textbook of entomology* (2nd edn). John Wiley and Sons, New York.

Ross, H. H. (1974). *Biological systematics*. Addison-Wesley, Reading, PA.

Saporta, G. de (1877). L'Ancienne vegetation polaire. *Comptes Rendus des Congres Nationales de Science Geographique*.

Sclater, P. L. (1858). On the general geographical distribution of the members of the class Aves. *Journal of the Linnean Society of London, Zoology* **2**, 130-145.

Shields, O. (1979). Evidence for the initial opening of the Pacific Ocean in the Jurassic. *Palaeogeography, Palaeoclimatography, Palaeoecology* **26**, 181-220.

Shields, O. (1983). Trans-Pacific links that suggest earth expansion. In *Expanding earth symposium* (ed. S. W. Carey) pp. 199-205. University of Tasmania.

Simpson, G. G. (1953). *The major features of evolution*. Columbia University Press, New York.

Simpson, G. G. (1962). *Evolution and geography*. Oregon State System of Higher Education, Eugene, OR.

Simpson, G. G. (1965). *The geography of evolution*. Chilton, Philadelphia.

Simpson, G. G. (1980). *Splendid isolation: the curious history of South American mammals*. Yale University Press, New Haven.

Smith, A. C. (1970). *The Pacific as a key to flowering plants history. Harold C. Lyon arboretum lecture number one*. University of Hawaii, Honolulu.

Smith, A. G., Briden, J. C., and Drewry, G.E. (1973). Phanerozoic world maps. In *Organisms and continents through time* (ed. N. F. Hughes). pp. 1-42. Palaeontological Association, London.

Stebbins, G. L. (1974). *Flowering plants: evolution above the species level*. Edward Arnold, London.

Steenis, C. G. G. J. van. (1953). Results of the Archbold expeditions; papuan *Nothofagus*. *Journal of the Arnold Arboretum, Harvard University* **34**, 301-374

Steenis, C. G. G. J. van (1962). The mountain flora of the Malaysian tropics. *Endeavour* **21**, 183-192.

Stott, P. (1981). *Historical plant geography*. George Allen and Unwin, London.

Sykes, L. R., McCann, W. R., and Kafka, L. (1982). Motion of Caribbean plate during last 7 million years and implications for earlier Cenozoic movements. *Journal of Geophysical Research* **87**, 10656-10676.

Takhtajan, A. (1969). *Flowering plants: origin and dispersal*. Oliver and Boyd, Edinburgh.

Tedford, R. H. (1981). Discussion (of A. Nur and Z. Ben Auraham, Lost Pacifica Continent: a mobilistic speculation). In *Vicariance biogeography: a critique* (eds G. Nelson and D. E. Rosen) pp. 367-370. Columbia University Press, New York.

Thorne, R. F. (1972). Major disjunctions in the geographic ranges of seed plants. *Quarterly Review of Biology* **47**, 365-411.

Tozer, E. T. (1982). Marine Triassic faunas of North America: their significance for assessing plate and terrane movements. *Geologische Rundschau* **71**, 1077-1104.

Turrill, W. B. (1953). *Pioneer plant geography. The phytogeographical researches of Sir Joseph Dalton Hooker*. Martinus Nijhoff, The Hague.

Wagner, W. H. Jr. (1961). Problems in the classification of ferns. In: *Recent Advances in Botany* (ed. Anon) pp. 841-844. University of Toronto Press, Montreal.

Wagner, W. H. Jr. (1980). Origin and philosophy of the groundplan-divergence method of cladistics. *Systematic Botany* **5**, 173-193.

Wallace, A. R. (1876). *The geographical distribution of animals*, (2 vols). Macmillan, London.

Wallace, A. R. (1880). *Island life: or the phenomena and causes of insular faunas and floras including a revision and attempted solution of the problem of geological climates.* Macmillan, London.

Webb, S. D. (1976). Mammalian faunal dynamics of the Great American Interchange. *Paleobiology* **2**, 216-234.

Webb, S. D. (1978). Mammalian faunal dynamics of the Great American Interchange: reply to an alternative interpretation. *Paleobiology* **4**, 206-209.

Wegener, A. (1915). *Die Enstehung der Kontinente und Ozeane. Sammlung Vieweg, nr. 23*. F. Vieweg und Sohn, Braunschweig.

Wildeman, E. de (1905). *Expedition antarctique Belge. Resultats du voyage du S. Y. Belgica. Botanique* Vol. 6 phanerogramés (pp. 222 + 23 plates). J. E. Buschmann, Anvers.

Wiley, E. O. (1980). Phylogenetic systematics and vicariance biogeography. *Systematic Botany* **5** (2), 194-220.

Wiley, E. O. (1981). *Phylogenetics: the theory and practice of phylogenetic systematics*. John Wiley and Sons, New York.

Willdenow, C. (1798). *Grundriss der Kräuterkunde* (2nd edn). Haude und Spener, Vienna.

Willis, J. C. (1922). *Age and area. A study in geographical distribution and origin of species*. Cambridge University Press, Cambridge.

Wulff, E. V. (1950). *An introduction to historical plant geography* (translated from the Russian by E. Brissenden). Chronica Botanica, Waltham, MA.

Zeil, W. (1979). *The Andes, a geological review*. Gebrüder Borntraeger, Berlin-Stuttgart.

GLOSSARY

amphitropical (antitropical, bipolar) A distribution characterized by taxa present in the northern and southern hemispheres, in particular the boreal and austral zones, but absent from the tropics.

analogue, analogy (cf. homologue) Used of similar characters or character states that have different but parallel modifications from other conditions, e.g. wings of birds and bats.

apomorphy Derived from (and different from) a generalized condition; used of characters, e.g. apomorphic characters.

area A biogeographic region occupied by a monophyletic group of organisms or a species.

austral zone Temperate regions of the southern hemisphere.

biogeography The study of what organisms live where on earth and why.

biota All the plants and animals occupying a given location.

boreal zone Temperate regions of the northern hemisphere.

centre of origin In dispersal biogeography, the area in which a taxon is supposed to have evolved.

cladistic biogeography The combination of cladistics (phylogenetic systematics) with vicariance biogeography. A method that searches for patterns of relationship among areas of endemism.

cladistics (phylogenetic systematics) A method of phylogeny reconstruction concerned with branching structure. Sister group relationships are determined on the basis of shared derived (apomorphic) characters.

cladogram A branching diagram specifying the hierarchical relationships of taxa, in which the taxa occupy the terminal nodes and the internal nodes are defined by apomorphic (shared derived) characters.

components Elements of a group of areas, or group of taxa, as determined by the branching points (nodes) of a cladogram, e.g. in a group comprising three taxa A, B, and C, when B is more closely related to C, there are two components—an ABC component and a BC component.

concordance The degree of agreement between patterns.

congruence Agreement in the systematic distribution of characters or topology of cladograms.

consensus tree A branching diagram derived from adding two or more cladograms together.

dispersal The movement of an organism from one area to another independent of other organisms and of earth history, that changes the natural distribution of the organism.

dispersion The movement of organisms and their offspring within an area.

ecological biogeography Spatial patterns of organisms considered in terms of interactions with other organisms and their environment.

endemic A taxon which is restricted to a given area and found nowhere else in the world.

generalized track Two or more tracks connecting the same two or more areas of endemism (see also track).

historical biogeography Spatial patterns of organisms interpreted in terms of their concordance to patterns of earth history.

homologue, homology, (cf. analogue) Used of similar characters or character states that share modifications from another condition, e.g. wings of birds in relation to forelimbs of other tetrapods.

Laurasia One of two supercontinents formed by the break up of the ancient supercontinent Pangaea; comprising North America, Greenland, Asia, and Europe (but not including the Indian subcontinent).

monophyly (monophyletic group) Derived from a common ancestor; used of a group in which all members are defined by the unique characters of the presumed ancestor.

Pacifica An ancient supercontinental land mass composed of bits of terrain now distributed around the Pacific margin. The theory is not accepted generally by geologists and biologists and details have not been worked out fully.

panbiogeography The examination of distribution and relationships on a worldwide scale; term first used by Croizat (1958).

Pangaea An ancient supercontinent comprising all known continental land masses, that began to break up during the Triassic.

paraphyly (paraphyletic group) A category or false

group based on the common possession of plesiomorphic characters (symplesiomorphy); a group which does not contain all of the descendants of a common ancestor.

pattern A set of relationships among taxa or areas as specified by a cladogram.

plate tectonics (including continental drift) Concept of earth's history that the earth's crust is composed of plates that move relative to each other. As continents are found on plates, *continental drift* is one result of this movement.

plesiomorphy A generalized character or condition shared by all members of a group.

progression rule The concept which predicts that the most plesiomorphic members of a taxon occupy the taxon's centre of origin, whereas the more apomorphic members have dispersed away from the centre, such that the more apomorphic a taxon, the farther from the centre it will occur.

sister taxa (areas) Two taxa (areas) that are more closely related to each other than either is to a third taxon (area).

synapomorphy A derived character, or character state, shared by and defining a group of organisms within the context of a large group.

track A graph or line which connects two or more areas of endemism.

transformation series A series of (three or more) increasingly apomorphic characters or character states.

vicariance The existence of closely related taxa or biota in disjunct areas, which have been separated by the formation of a natural barrier (vicariance event).

vicariance biogeography The study of relationships of biota in different geographical areas as shown by similar patterns in unrelated groups.

vicariance event The splitting of a taxon or biota into two or more geographical subdivisions by the formation of natural barriers e.g. mountains, tectonic events, rivers.

INDEX OF ANIMAL AND PLANT NAMES

Acaena 1, 2, *6*, 8
African Violets 84
Agoseris 73
Algae 24
Amphibians 29
Amsinckia 73
Angiosperms 15, 24
Araucaria 19, 69
Armeria 73
Atydid shrimps *20*

Bahia 73
bats 23, 58
beeches 69, 74, 76, *79*, 82, 83, 87
beetles 83
  (Broscini) 69
  (Byrrhinae) 69
  (Derodontidae) 69
  (Nemonychidae) 69
Bignoniaceae 71
birds 18, 23, 29, 38, 40–3, *40*, *41*, *43*, 60
bowfins 61, *62*
brachyuran crabs *20*
brown algae *22*
Bryophyta *22*
buttercups 76

caddis flies 26, 84
carabid beetles 69
carrot family 1
Ceratophyllaceae *3*
Ceratophyllum 2
  C. demersum 1, *3*
Chaleosyrphus 8
characin fishes 55, 59
characiform fishes *56*
Charales *22*
Charophyta *22*
Chironomidae *28*
  (tribe Boreoheptagyini) 28
  (subfamily Diamesinae) 27, *28*, 74
  (tribe Diamesinae) *28*
  (tribe Diamesini) *28*
  (tribe Harrisonini) *28*
  (subfamily Heptagyini) 27
  (tribe Heptagyini) *28*
  (tribe Lobodiamesini) 28, *28*
  (subfamily Podonominae) 74
  (tribe Protanypodini) *28*
chironomid midges 19, 26, 27, 74
Chironomas plumosus *74*
cichlids 71, 76, 83
Clarkia 73
clubmosses *22*
Coleoptera (see beetle tribes) 69
Colombophiloscia *20*
Colochaete *22*
Compositae *2*
conifers *22*, 24
conjugates *22*
Coranthera *86*
Coriaria 8, 76
crayfish 77
crowberries 69
cycads *22*, 24

daisies 21
Dallia 61
dandelion 1
Depanthus *86*
devil's coach horse beetles 69
diamesine midges *74*, *79*
didelphids 57, *58*, 60
Didelphis 55, *56*
  D. azarae 57
Dilleniales 11
Diospyros 71, *72*
Diptera *28*
duck-billed platypus 23
duckweed 10

earthworms 18
ebonies 71, 83
edentates 58
Embryobionta *22*
Empetraceae *7*
Empetrum 3, *7*, 69, 73
  E. nigrum 3
  E. rubrum 3
Ephedra *22*, 24
Eucalyptus 54
Euphrasia 69, 79, 80, 83
eutherians 58
eyebright 69

Fagaceae 3, 10, 69, 73
Fagus 73, *74*, *79*, 87
ferns *22*, 38, *40*, 42, *43*, 46, *46*, *49*
Fieldia *86*
fish 29, 32–3, *32*, 38, *40*, 42, *43*, 46, *47–9*, 60
  (Percichthyidae) 69, *72*
flowering plants 18, 21, *22*
flowering trees 32, *32*, 33
freshwater fishes 18, 61
frogs 38, 40, 41
  (Microhylinae) 76
foxes 21

galaxiid fishes 69
galliform birds 79, *79*
Garnieria *86*
Gesneriaceae 84, *86*
Gevuina 8
Gilia 73
ginkgo *22*
Gnetum *22*, 24
Gobiidae 2, *6*
green algae *22*
Gunnera 76
gymnosine staphylinids 69
Gymnospermae *22*, 24, 25
gymnosperms 24

Haplopappus 73
Hebe 76
Heptagyia 28, *28*
Heterandria 33, 35, 37–8, *35–40*, 47–50, *50–1*
  H. anzuetoi *36*, 38
  H. attenuata *36*, 38
  H. bimaculata *36*, 38
  H. cataractae *34*, *36*, 38
  H. dirempta *36*, 38
  H. formosa *36*
  H. jonesi *36*, 38
  H. litoperas *36*, 38
  H. obliqua *36*
heterandriine killifishes 55
heterandriine poeciliid fishes *56*
Hippurus 73
hoop pine 69
hornwarts *22*
horsetails *22*
Hydrobiosella 84, *86*
hylid frogs 8, 79, *79*

Indian bean tree 71
insects 18
Invertebrata 24
isopods *20*
ivory nut palms 31

Jimenezia *20*

Klebsorbium *22*
Killifishes 33, 59, 71, 76, 79, 80, 82, 83
Koenigia 73

lampreys 69, 83
land plants 21
Libocedrus *19*
Limaya 28, *28*
lizards 23, 38, 40, *40*, 41
Liverworts *22*
Lomatia 8

Magnolia 55
magnolias *56*
Magnoliales 11
Magnoliatae 25
Magnoliophyta *22*
mammals 8, 9, 18, 21, 23, 29, 55, 58–60
Maoridiamesa 28, *28*
marsupials 8, *8*, 58, 59, 79, 80
Megalomys *20*
midges 69, 74, 76, 83, 87
mole 76
molluscs 18
monkey puzzle 69
monotremes 23
mordaciid lampreys 69
mosses 18, *22*
moths 38, 40–2, *40*, *41*, *43*
mud-minnows 61
multituberculates 58
Myristicaceae 71

nandids 71
Nesophiloscia *20*
Nicotiana 8
nutmegs 71
Nothofagus 3, 8, 10, 19, *19*, 69, 70, 73, *74*, 78, *79*, 83–4, *86*, 87
  N. recurva *71*
northern crowberry 3

Orites 8
Osmorhiza 1, 3
  O. chilensis 1, 3, *5*
Orestias 79, 80
Oreomyrrhis 76
osteoglosine fishes 8
Oreocallis 8

Paraheptgyia 28, *28*
palaearctic mammals 14
paddlefishes 61, *62*
percichthyid fishes 81, 83
percomorph fish *72*
Persoonia *86*
Persooninae 84, *86*
Phippsia 73
Phytelephas 31
pigs 23
Pinophyta *22*
plants 29
platycercine parrots 84, *86*
platy-fishes 33
Plectritis 73
plethodontid salamanders 71
podonomine 74, *79*
poeciliid fish 33, 47, 52
Primula 73
Proteaceae 8
Pro-gymnospermopsida 25
psilophytes *22*
Pyramimonas *22*

ratite birds 8, 79, *79*
rats 21, 23
red algae *22*
Reissia 28, *28*
reptiles 29
Reptilia 24, 25
Restionaceae *19*
Rhabdothamnus *86*
rhyniophytes *22*
rice rats *20*
rivulid killifishes 55, *56*
rivulids 60
Rosaceae 1, *6*

Scheuchzeria palustris 71
Schedonnardus 73
sheep 23
silverside fishes 69
snakes
  (Colubridae) 77
  (Pareinae) 77
  (Xenoderminae) 76
southern beech 3, 19, *70*, 73, 78, 79
southern crowberry 3
southern speedwells 76
stomatopod crustaceans *20*
Stylidiaceae 8
sunflowers 21
sungrebes 71
swordtails 33
syrphid flies 8

Taraxacum magellanicum 1, *2*
tetrapods 21
Thallobionta *22*
Toronia *86*
tree frogs 76
trees 38, *40*, 42, *43*, *45*, 46, *48*, *49*
Troglophiloscia *20*
Typhlataya *20*

Umbelliferae 1
umbrids 61, *62*, 63
Urotrichus 76

waterweed hornwort 1
Welwitschia *22*, 24
Winteraceae 3, 8
Wormaldia *27*
  W. kisoensis 26, *27*
  W. mohri *27*
worms 38, 40–2, *40*, *41*, *43*

Xiphophorus 33, 35, *35–40*, 37, 38, 47–50, *50*, *51*
  X. alvarezi *36*, 47
  X. clemenciae *36*, 38
  X. couchianus *36*
  X. corteze *34*, *36*, 38
  X. evelynae 36
  X. gordoni *31*
  X. helleri *36*, 38
  X. maculatus *36*
  X. milleri *36*
  X. montezumae *36*, 38
  X. nigrensis *36*, 38
  X. 'PMH' *36*
  X. pygmaeus *36*, 38
  X. signum *36*, 38
  X. variatus *36*
  X. xiphidium *36*
xylotine sawflies 79, *79*

GENERAL INDEX

'accreted terranes' 63
Adams 14
Africa 8, 12, 32, 40–6, 69, 75
Age of Fish 58
Age of Mammals 58
Age of Reptiles 58
Alaska 61, 63, 75
alpha taxonomy xii
amphi-American marine biota 19, 20
amphitropical distributions 69–73, 76–7, 84
analytical biogeography xii
analytical taxonomy xii
ancestral cosmopolitanism 73
Andean South America 28
Andes 2, 76, 81
Antarctic 11
Antarctica 2, 14, 28, 69
Arctic 3, 76
areas of endemism xii, 1, 10, 12, 16, 18–21, 26, 32, 38, 41, 50 53, 57, 64–6, 78
Argentina 81
Asia 8, 15, 27, 61–4, 74–6, 80, 82
Asian Pacific 26
Atlantic 75–6, 79–81
Australia 3, 11, 18, 28, 32, 38, 38–46, 69, 73, 74, 77–9, 84
Australasia 15, 54, 77
autapomorphies 23, 24
austral zone 69–74, 78–86

Baja California 63
Bering Strait 14, 27
'biphyletic' nature of the North American biota 61–4
bipolar distributions 69
Bolivia 18, 81
Bonpland 9
boreal zone 70–3, 77–81, 84–6
British Isles 10, 11, 13
Brundin 26, 28

Cain 14, 19
California 2, 8, 54, 75–6
Caribbean 1, 29–30, 52, 55, 60, 76
Caribbean biogeography, the application of Croizat's vicariance method 29–31
Central Andes 74, 80–2, 87
Central America 52
Central American archipelago 30
Central North America 61
'centre of origin' xii, 7, 9, 10, 14, 15, 18, 25, 26, 77
centres of origin, Cain's criteria for recognition 14
Chile 76, 81–2
Colombia 76
Colorado 63
component analysis of cladograms 38–52
congruence 67
continental drift, theory of 21, 58, 71, 75, 82
cosmopolitan species 10, 12
cradle of the flowering plants 14, 15
Craw 84
Croizat 1, 16, 17, 19, 21, 26, 29, 31

Darlington 71–8
Darwin 1, 10–12, 19–22
Darwin–Wallace tradition 12, 15–17
de Buffon 8, 10, 12, 17, 18
de Candolle 1, 9–12, 17, 19
dispersal biogeography 1
Du Reitz 73
dynamics of community structure xi

East Asia 77
eastern Atlantic, the 29
Eastern Australia 82
eastern Cordilleras of the continental Andes 81
eastern Honduras 33
eastern Nicaragua 35
eastern North America 1, 26–7, 30, 61–4, 76, 80
eastern Pacific, the 29–30
eastern Siberia 61
East Pacific rise 75
Ecuador 81
episodic growth of continents 75
Eurasia 3, 14, 73
Europe 11, 14, 57, 61–4, 74–80
'expanding' earth theory 75, 81, 83, 84

factors in present day distribution of plants and animals 54
Faunal realms 13, *13*, 17
Forbes 10, 12
'form making' of taxa 17

Galapagos 11
generalized tracks 18, 19, 20, *20*, 29–32, 37
glacial pathways 54
Gondwanaland 15
Good 52
'Great American Interchange' 59
Gulf of Fonseca 52

'habitations' 9
Hawaii 2, 18
Hennig 21, 23, 24, 26
Hennig's concepts of character distribution 23
Hennigian systematics 21
Hooker 11, 12, 17, 84
Humboldt 1, 8, 9, 19, 84
Humphries 78

incongruence in cladograms 38
incongruent distribution patterns 42
influence of external elements on plant geography 9
'isolating dispersal' 12
isthmian America 52
isthmus of Panama 19

Japan 26, 27, 76

Laurasia 28
Linnaeus 1, 7, 19, 21
Lyell 12
Lyell's principle 11

Madagascar 77
Magellanian Pacifica 82
Malfait & Dinkelman's theory 30
Mediterranean 76
Melanesia 15
Mesoamerica 29, 50, 51
Mexico 8, 18, 33
microplate accretions along the Gondwanian margin 84
Middle America 1, 33, 54, 55, 59, 60
monophyletic groups 23–6
mutability of species, theory of 11

narrative biogeography xii
narrative taxonomy xii
Nelson 26
Nelson & Platnik 52
New Caledonia 3, 14, 69, 75, 78, 81, 84
New Guinea 3, 32–3, 38–45, 69, 75, 78, 80, 84
New World 54–5, 60

New Zealand 3, 14, 18, 27–8, 32, 69, 73–8, 80, 83–7
New Zealand, early separation from Western Antarctica 28
Nicaragua 52
north America 27, 79
North America 1, 3, 14, 27–30, 52–60, 63–6, 73–9
northern Argentina 76
Northern Canada 1
north-eastern North America 71
northern North America 71
north-west South America 31

Owen 76

Pacifica, break up of 75–7, 81, 84–7
Pacific, the 15, 29, 75–6, 79–82
Panamanian land bridge 55
Pangaea, break up of 75–7, 81–7
Papua New Guinea 32
paraphyletic groups 23–5
Patagonia 28, 78, 80, 87
Patagonian Andes 73–4
Peru 18, 82
Peruvian Pacifica 82
phylogenetic biogeography xii
plate tectonics 75, 77, 83
polyphyletic groups 23
'pre-Pangaea' 86
primordial island, the 7
principle of parsimony 83
proto-Antillean archipelago 29

Queensland 14

Raven and Axelrod 78
'reciprocal illuminators' 66
redundant, missing, and ambiguous information in cladograms 38
repetition of distribution patterns 28
Rocky mountain system 52
Rosen 26, 29, 30, 32, 35, 37, 48
Ross 26

Sarawak 26
Sclater 13
'shared environmental effects' 54
Simpson 1, 19
Smoky mountains 26
South Africa 3, 28, 33, 74, 80
South America xi, 1, 2, 8, 12, 14, 28–32, 40–6, 52–60, 73–87
South-east Asia xi, 14, 76, 77
south-eastern Australia 28
south-eastern North America 61
Southern Africa 27
southern Andes 81–2
southern South America 3, 32, 38, 69
'specific centres' 10
sympatric occurrence 3
symplesiomorphies 23–4
synapomorphies 23–4
syngameons 73

Takhtajan 15
Tasmania 3, 11, 18, 27, 69, 73, 78, 80, 84
Tasman sea 84
temperate southern America 55
Thorne 73
track anlaysis 52
tropical Africa 78, 79

Utah 63

'vacuum' theory of biogeography 26
Venezuela 81
vicariance biogeography xii
vicariance event 67

Wallace 1, 12, 13, 19
Washington State 63
western Atlantic, the 29
western cordilleras of North America 75
western North America 18, 30, 52, 61–4, 76, 80, 82
western Pacific, the 1, 15, 29–30, 75–6, 79
west Gondwanaland 77
Wiley 37, 38, 50
Willdenow 10, 19
Wulff 17
Wyoming 63

Yunnan 14